Lecture Notes in Physics
Monographs

Springer
Berlin
Heidelberg
New York
Barcelona
Hong Kong
London
Milan
Paris
Singapore
Tokyo

Physics and Astronomy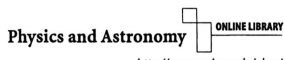

http://www.springer.de/phys/

The Editorial Policy for Monographs

The series Lecture Notes in Physics reports new developments in physical research and teaching - quickly, informally, and at a high level. The type of material considered for publication in the monograph Series includes monographs presenting original research or new angles in a classical field. The timeliness of a manuscript is more important than its form, which may be preliminary or tentative. Manuscripts should be reasonably self-contained. They will often present not only results of the author(s) but also related work by other people and will provide sufficient motivation, examples, and applications.

The manuscripts or a detailed description thereof should be submitted either to one of the series editors or to the managing editor. The proposal is then carefully refereed. A final decision concerning publication can often only be made on the basis of the complete manuscript, but otherwise the editors will try to make a preliminary decision as definite as they can on the basis of the available information.

Manuscripts should be no less than 100 and preferably no more than 400 pages in length. Final manuscripts should be in English. They should include a table of contents and an informative introduction accessible also to readers not particularly familiar with the topic treated. Authors are free to use the material in other publications. However, if extensive use is made elsewhere, the publisher should be informed. Authors receive jointly 30 complimentary copies of their book. They are entitled to purchase further copies of their book at a reduced rate. No reprints of individual contributions can be supplied. No royalty is paid on Lecture Notes in Physics volumes. Commitment to publish is made by letter of interest rather than by signing a formal contract. Springer-Verlag secures the copyright for each volume.

The Production Process

The books are hardbound, and quality paper appropriate to the needs of the author(s) is used. Publication time is about ten weeks. More than twenty years of experience guarantee authors the best possible service. To reach the goal of rapid publication at a low price the technique of photographic reproduction from a camera-ready manuscript was chosen. This process shifts the main responsibility for the technical quality considerably from the publisher to the author. We therefore urge all authors to observe very carefully our guidelines for the preparation of camera-ready manuscripts, which we will supply on request. This applies especially to the quality of figures and halftones submitted for publication. Figures should be submitted as originals or glossy prints, as very often Xerox copies are not suitable for reproduction. For the same reason, any writing within figures should not be smaller than 2.5 mm. It might be useful to look at some of the volumes already published or, especially if some atypical text is planned, to write to the Physics Editorial Department of Springer-Verlag direct. This avoids mistakes and time-consuming correspondence during the production period.

As a special service, we offer free of charge LaTeX and TeX macro packages to format the text according to Springer-Verlag's quality requirements. We strongly recommend authors to make use of this offer, as the result will be a book of considerably improved technical quality.

For further information please contact Springer-Verlag, Physics Editorial Department II, Tiergartenstrasse 17, D-69121 Heidelberg, Germany.

Series homepage – http://www.springer.de/phys/books/lnpm

Alexander S. Holevo

Statistical Structure of Quantum Theory

 Springer

Author

Alexander S. Holevo
Steklov Mathematical Institute
Gubkina 8
117966 Moscow, GSP-1, Russian Federation

Library of Congress Cataloging-in-Publication Data applied for.

Die Deutsche Bibliothek - CIP-Einheitsaufnahme

Cholevo, Aleksandr S.:
Statistical structure of quantum theory / Alexander S. Holevo. -
Berlin ; Heidelberg ; New York ; Barcelona ; Hong Kong ; London ;
Milan ; Paris ; Singapore ; Tokyo : Springer, 2001
 (Lecture notes in physics : N.s. M, Monographs ; 67)
 (Physics and astronomy online library)
 ISBN 3-540-42082-7

ISSN 0940-7677 (Lecture Notes in Physics. Monographs)
ISBN 3-540-42082-7 Springer-Verlag Berlin Heidelberg New York

Springer-Verlag Berlin Heidelberg New York
a member of BertelsmannSpringer Science+Business Media GmbH

http://www.springer.de

Typesetting: Camera-ready copy from the author
Cover design: *design & production*, Heidelberg

Printed on acid-free paper
SPIN: 10720547   55/3141/du - 5 4 3 2 1 0

Preface

During the last three decades the mathematical foundations of quantum mechanics, related to the theory of quantum measurement, have undergone profound changes. A broader comprehension of these changes is now developing.

It is well understood that quantum mechanics is not a merely dynamical theory; supplied with the statistical interpretation, it provides a new kind of probabilistic model, radically different from the classical one. The statistical structure of quantum mechanics is thus a subject deserving special investigation, to a great extent different from the standard contents of textbook quantum mechanics.

The first systematic investigation of the probabilistic structure of quantum theory originated in the well-known monograph of J. von Neumann "Mathematical Foundations of Quantum Mechanics" (1932). A year later another famous book appeared – A. N. Kolmogorov's treatise on the mathematical foundations of probability theory. The role and the impact of these books were significantly different. Kolmogorov completed the long period of creation of a conceptual basis for probability theory, providing it with classical clarity, definiteness and transparency. The book of von Neumann was written soon after the birth of quantum physics and was one of the first attempts to understand its mathematical structure in connection with its statistical interpretation. It raised a number of fundamental issues, not all of which could be given a conceptually satisfactory solution at that time, and it served as a source of inspiration or a starting point for subsequent investigations.

In the 1930s the interests of physicists shifted to quantum field theory and high-energy physics, while the basics of quantum theory were left in a rather unexplored state. This was a natural process of extensive development; intensive exploration required instruments which had only been prepared at that time. The birth of quantum mechanics stimulated the development of an adequate mathematical apparatus – the operator theory – which acquired its modern shape by the 1960s.

By the same time, the emergence of applied quantum physics, such as quantum optics, quantum electronics, optical communications, as well as the development of high-precision experimental techniques, has put the issue of consistent quantitative quantum statistical theory in a more practical setting.

Such a theory was created in the 1970s–80s as a far-reaching logical extension
of the statistical interpretation, resting upon the mathematical foundation of
modern functional analysis. Rephrasing the well-known definition of probability theory[1], one may say that it is a theory of operators in a Hilbert space
given a soul by the statistical interpretation of quantum mechanics. Its mathematical essence is diverse aspects of positivity and tensor product structures
in operator algebras (having their roots, correspondingly, in the fundamental probabilistic properties of positivity and independence). Key notions are,
in particular, *resolution of identity* (positive operator valued measure) and
completely positive map, generalizing, correspondingly, spectral measure and
unitary evolution of the standard quantum mechanics.

The subject of this book is a survey of basic principles and results of
this theory. In our presentation we adhere to a pragmatic attitude, reducing to a minimum discussions of both axiomatic and epistemological issues
of quantum mechanics; instead we concentrate on the correct formulation
and solution of quite a few concrete problems (briefly reviewed in the Introduction), which appeared unsolvable or even untreatable in the standard
framework. Chapters 3–5 can be considered as a complement to and an extension of the author's monograph "Probabilistic and Statistical Aspects of
Quantum Theory", in the direction of problems involving state changes and
measurement dynamics[2]. However, unlike that book, they give a concise survey rather than a detailed presentation of the relevant topics: there are more
motivations than proofs, which the interested reader can find in the refer-
ences. A full exposition of the material considered would require much more
space and more advanced mathematics.

We also would like to mention that a finite dimensional version of this
generalized statistical framework appears to be an adequate background for
the recent investigations in quantum information and computing, but these
important issues require separate treatment and were only partially touched
upon here.

This text was completed when the author was visiting Phillips-Universität
Marburg with the support of DFG, the University of Milan with the support
of the Landau Network–Centro Volta, and Technische Universität Braun-
schweig with A. von Humboldt Research Award.

Moscow, April 2001

A.S. Holevo

---

[1] "Probability theory is a measure theory – with a soul" (M. Kac).
[2] This text is the outcome of an iterative process, the first approximation for which
was the author's survey [119], translated into English by Geraldine and Robin
Hudsons. The LaTeX files were created with the invaluable help of J. Plumbaum
(Phillips-Universität Marburg) and S.V. Klimenko (IHEP, Protvino).

Contents

0. Introduction 1
   0.1 Finite Dimensional Systems 1
   0.2 General Postulates of Statistical Description 3
   0.3 Classical and Quantum Systems 4
   0.4 Randomization in Classical and Quantum Statistics 5
   0.5 Convex Geometry and Fundamental Limits
       for Quantum Measurements 6
   0.6 The Correspondence Problem 8
   0.7 Repeated and Continuous Measurements 8
   0.8 Irreversible Dynamics 10
   0.9 Quantum Stochastic Processes 11

1. The Standard Statistical Model
   of Quantum Mechanics 13
   1.1 Basic Concepts 13
       1.1.1 Operators in Hilbert Space 13
       1.1.2 Density Operators 15
       1.1.3 The Spectral Measure 16
       1.1.4 The Statistical Postulate 17
       1.1.5 Compatible Observables 18
       1.1.6 The Simplest Quantum System 20
   1.2 Symmetries, Kinematics, Dynamics 22
       1.2.1 Groups of Symmetries 22
       1.2.2 One-Parameter Groups........................... 23
       1.2.3 The Weyl Relations 24
       1.2.4 Gaussian States 27
   1.3 Composite Systems 29
       1.3.1 The Tensor Product of Hilbert Spaces............. 29
       1.3.2 Product States 30
   1.4 The Problem of Hidden Variables........................ 31
       1.4.1 Hidden Variables and Quantum Complementarity 31
       1.4.2 Hidden Variables and Quantum Nonseparability 33
       1.4.3 The Structure of the Set of Quantum Correlations 36

2. **Statistics of Quantum Measurements** 39
   2.1 Generalized Observables 39
      2.1.1 Resolutions of the Identity 39
      2.1.2 The Generalized Statistical Model
             of Quantum Mechanics 41
      2.1.3 The Geometry of the Set of Generalized Observables .. 43
   2.2 Quantum Statistical Decision Theory 45
      2.2.1 Optimal Detection 45
      2.2.2 The Bayes Problem 47
      2.2.3 The Quantum Coding Theorem 50
      2.2.4 General Formulation 53
      2.2.5 The Quantum Cramér–Rao Inequalities 56
      2.2.6 Recent Progress in State Estimation 59
   2.3 Covariant Observables 61
      2.3.1 Formulation of the Problem 61
      2.3.2 The Structure of a Covariant Resolution of the Identity 62
      2.3.3 Generalized Imprimitivity Systems 64
      2.3.4 The Case of an Abelian Group 65
      2.3.5 Canonical Conjugacy in Quantum Mechanics 67
      2.3.6 Localizability 70

3. **Evolution of an Open System** 71
   3.1 Transformations of Quantum States
        and Observables 71
      3.1.1 Completely Positive Maps 71
      3.1.2 Operations, Dynamical Maps 73
      3.1.3 Expectations 75
   3.2 Quantum Channel Capacities 76
      3.2.1 The Notion of Channel 76
      3.2.2 The Variety of Classical Capacities 77
      3.2.3 Quantum Mutual Information 79
      3.2.4 The Quantum Capacity 80
   3.3 Quantum Dynamical Semigroups 81
      3.3.1 Definition and Examples 81
      3.3.2 The Generator 83
      3.3.3 Unbounded Generators 85
      3.3.4 Covariant Evolutions 88
      3.3.5 Ergodic Properties 91
      3.3.6 Dilations of Dynamical Semigroups 93

4. **Repeated and Continuous
   Measurement Processes** 97
   4.1 Statistics of Repeated Measurements 97
      4.1.1 The Concept of Instrument 97
      4.1.2 Representation of a Completely Positive Instrument .. 100

    4.1.3   Three Levels of Description of Quantum Measurements 102
    4.1.4   Repeatability 103
    4.1.5   Measurements of Continuous Observables 105
    4.1.6   The Standard Quantum Limit 106
  4.2   Continuous Measurement Processes 108
    4.2.1   Nondemolition Measurements 108
    4.2.2   The Quantum Zeno Paradox...................... 109
    4.2.3   The Limit Theorem for Convolutions of Instruments .. 111
    4.2.4   Convolution Semigroups of Instruments 113
    4.2.5   Instrumental Processes........................... 115

5. Processes in Fock Space 119
  5.1   Quantum Stochastic Calculus 119
    5.1.1   Basic Definitions 119
    5.1.2   The Stochastic Integral 121
    5.1.3   The Quantum Ito Formula 124
    5.1.4   Quantum Stochastic Differential Equations 126
  5.2   Dilations in the Fock Space 130
    5.2.1   Wiener and Poisson Processes in the Fock Space 130
    5.2.2   Stochastic Evolutions and Dilations
            of Dynamical Semigroups........................ 133
    5.2.3   Dilations of Instrumental Processes 136
    5.2.4   Stochastic Representations
            of Continuous Measurement Processes 138
    5.2.5   Nonlinear Stochastic Equations of Posterior Dynamics 140

Bibliography.. 145

Index 157

0. Introduction

0.1 Finite Dimensional Systems

Exposition of probability theory usually begins with a finite scheme. Following this tradition, consider a finite probabilistic space $\Omega := \{1, \ldots, N\}$. Three closely interrelated facts express in different ways the classical character of the probabilistic scheme:

1. the set of events $A \subset \Omega$ forms a Boolean algebra,
2. the set of probability distributions $P = [p_1, \ldots, p_N]$ on Ω is a simplex, that is a convex set in which each point is uniquely expressible as a mixture (convex linear combination) of extreme points,
3. the set of random variables $X = [\lambda_1, \ldots, \lambda_N]$ on Ω forms a commutative algebra (under pointwise multiplication).

The quantum analogue of this scheme is the model of a N-level system. The analogue of probability distribution - the *state* of such a system - is described by a *density matrix*, that is a $N \times N$ Hermitian matrix S, which is positive definite and has unit trace:

$$S \geq 0, \quad \operatorname{Tr} S = 1. \tag{0.1}$$

The analogue of random variable - an *observable* - is described by an arbitrary $N \times N$ Hermitian matrix X. Let

$$X = \sum_{j=1}^{n} x_j E_j \tag{0.2}$$

be the spectral representation of the Hermitian matrix X, where $x_1 < x_2 < \cdots < x_n$ are the eigenvalues and $E_1, E_2, \ldots, E_n$ are the projectors onto the corresponding eigenspaces. The family $E = \{E_j; j = 1, \ldots, n\}$ forms an orthogonal resolution of identity:

$$E_j E_k = \delta_{jk} E_j, \quad \sum_{j=1}^{n} E_j = I, \tag{0.3}$$

where I is the unit matrix. From the properties (0.1) and (0.3) it follows that the relation

$$\mu_S^X(x_j) := \operatorname{Tr} SE_j \text{ for } j = 1, \ldots, n, \tag{0.4}$$

defines a probability distribution on the spectrum $\operatorname{Sp} X = \{x_1, \ldots, x_n\}$ of the observable X. In quantum mechanics it is postulated that this is the probability distribution of the observable X in the state S. In particular the mean value of X in the state S is equal to

$$\mathbf{E}_S(X) = \operatorname{Tr} SX. \tag{0.5}$$

If in this model we consider only diagonal matrices

$$S = \begin{pmatrix} p_1 & & 0 \\ & \ddots & \\ 0 & & p_N \end{pmatrix}, \quad X = \begin{pmatrix} \lambda_1 & & 0 \\ & \ddots & \\ 0 & & \lambda_N \end{pmatrix}$$

then we return to the classical scheme with N elementary events, where, in particular (0.5) reduces to $\mathbf{E}_S(X) = \sum_{j=1}^{N} p_j \lambda_j$. We obtain the same by considering only simultaneously diagonalizable (that is commuting) matrices. Since there are observables described by noncommuting matrices, the model of N-level system does not reduce to this classical scheme.

The role of indicators of events in the quantum case is played by observables having the values 0 or 1, that is by idempotent Hermitian matrices, $E^2 = E$. Introducing the unitary coordinate space $\mathcal{H} = \mathbb{C}^N$ in which the $N \times N$- matrices operate, we may consider such a matrix E as an orthogonal projector onto a subspace $\mathcal{E}$ in $\mathcal{H}$. In this way, *quantum events* can be identified with subspaces of the unitary space $\mathcal{H}$. The set of quantum events, called the *quantum logic*, is partially ordered (by the inclusion) and possesses the operations $\mathcal{E}_1 \vee \mathcal{E}_2$ (defined as the linear hull $\operatorname{Lin}(\mathcal{E}_1, \mathcal{E}_2)$ of the subspaces $\mathcal{E}_1, \mathcal{E}_2 \subset \mathcal{H}$), $\mathcal{E}_1 \wedge \mathcal{E}_2$ (defined as the intersection $\mathcal{E}_1 \cap \mathcal{E}_2$ of the subspaces $\mathcal{E}_1, \mathcal{E}_2 \subset \mathcal{H}$) and $\mathcal{E}'$ (defined as the orthogonal complement $\mathcal{E}^\perp$ of $\mathcal{E} \subset \mathcal{H}$) with the well known properties. The non-classical character of the model of N-level system can be expressed in three different ways:

1. the quantum logic of events is not a Boolean algebra; we do not have the distributivity law

$$\mathcal{E}_1 \wedge (\mathcal{E}_2 \vee \mathcal{E}_3) = (\mathcal{E}_1 \wedge \mathcal{E}_2) \vee (\mathcal{E}_1 \wedge \mathcal{E}_3)$$

   does not hold. Consequently there are no "elementary events" into which an arbitrary quantum event could be decomposed;
2. the convex set of states is not a simplex, that is, the representation of a density matrix as a mixture of extreme points is non-unique;
3. the complex linear hull of the set of observables is a non-commutative (associative) algebra.

In the infinite-dimensional case, instead of the matrices one has to consider operators acting in a certain Hilbert space $\mathcal{H}$. A mathematical exposition of the fundamental concepts of quantum mechanics in Hilbert space was first given by J. von Neumann [175]. In particular he emphasized the distinction between Hermitian (symmetric) and selfadjoint operators, which, of course, had not occured in the preceding physicists' works, and pointed out that it is in fact the condition of selfadjointness, which in the case of infinite dimensional Hilbert spaces underlies an analogue of the spectral decomposition (0.2). Another circle of problems in the case of an arbitrary Hilbert space is associated with the concept of trace and the corresponding class of operators with finite trace. This mathematical scheme, called the *standard statistical model* of quantum mechanics, will be reviewed in the Chap. 1.

0.2 General Postulates of Statistical Description

Each of the mathematical structures - the quantum logic of events, the convex set of states and the algebra of quantum observables - can be characterized by certain system of axioms, but the characterization problems arising in this way are highly nontrivial and constitute essentially separate research field, a review of which goes beyond the scope of the present notes (see, in particular [206], [162], [66], [71], [161], [222]). Since we are dealing with a concrete object - the quantum theory in a Hilbert space - we do not have to resort to one or another type of axiomatic systems; furthermore, it is precisely the peculiar characteristics of this concrete object which provide the fundamental motivation for various axiomatic schemes. However, one useful lesson of the axiomatic approach is the revelation of the fruitful parallelism between statistical descriptions of classical and quantum systems. The axioms below are modifications of the first four axioms from [162] that are equally applicable to both classical and quantum systems.

Axiom 1 *Let there be a given set $\mathfrak{S}$ whose elements are called states and a set $\mathfrak{O}$ whose elements are called (real) observables. For arbitrary $S \in \mathfrak{S}$ and $X \in \mathfrak{O}$ there is a probability distribution μ_S^X on the σ-algebra $\mathcal{B}(\mathbb{R})$ of Borel subsets of the real line $\mathbb{R}$, called the probability distribution of the observable X in the state S.*

The state S is interpreted as a more or less detailed description of the preparation of a *statistical ensemble* of independent individual representatives of the system under consideration, and the observable X – as a quantity, which can be measured by a definite apparatus for each representative in the given ensemble. Axiom 1 thus presupposes the *reproducibility* of the individual experiments and the *stability of frequencies* under independent repetitions. The following axiom expresses the possibility of the *mixing* of ensembles.

Axiom 2 *For arbitrary $S_1, S_2 \in \mathfrak{S}$ and an arbitrary number p with $0 < p < 1$, there exists $S \in \mathfrak{S}$ such that $\mu_S^X = p\mu_{S_1}^X + (1-p)\mu_{S_2}^X$ for all $X \in \mathfrak{O}$. S is said to be a mixture of the states S_1 and S_2 in the proportion $p : (1-p)$.*

The following axiom describes the possibility of processing the information obtained from a measurement of observable. Let f be a Borel function from $\mathbb{R}$ to $\mathbb{R}$. If $X_1, X_2 \in \mathfrak{O}$ are such that for all $S \in \mathfrak{S}, B \in \mathcal{B}(\mathbb{R})$

$$\mu_S^{X_2}(B) = \mu_S^{X_1}\left(f^{-1}(B)\right)$$

where $f^{-1}(B) := \{x \in \mathbb{R} : f(x) \in B\}$, then the observable X_2 is *functionally subordinate* to the observable X_1. In this case we shall write $X_2 = f \circ X_1$.

Axiom 3 *For arbitrary $X_1 \in \mathfrak{O}$ and an arbitrary Borel function f there exists $X_2 \in \mathfrak{O}$, such that $X_2 = f \circ X_1$.*

A pair of non-empty sets $(\mathfrak{S}, \mathfrak{O})$ satisfying the axioms 1 - 3 is said to be a *statistical model*. The statistical model is said to be *separated* if the following axiom is valid.

Axiom 4 *From the fact that $\mu_{S_1}^X = \mu_{S_2}^X$ for all $X \in \mathfrak{O}$ it follows that $S_1 = S_2$ and from $\mu_S^{X_1} = \mu_S^{X_2}$ for all $S \in \mathfrak{S}$ it follows that $X_1 = X_2$.*

For a separated model both the operation of mixing in $\mathfrak{S}$ and the functional subordination in $\mathfrak{O}$ are uniquely defined. Thus the set of states $\mathfrak{S}$ obtains a convex structure and the set of observables $\mathfrak{O}$ – a partial ordering.

Observables $X_1, \ldots, X_m$ are called *compatible* if they are all functionally subordinate to some observable X, that is, $X_j = f_j \circ X$ for $j = 1, \ldots, m$. Compatible observables can be measured in a single experiment. The observables which are compatible with all observables $X \in \mathfrak{O}$ form the *center* or the *classical part* of the statistical model.

0.3 Classical and Quantum Systems

Let $(\Omega, \mathcal{B}(\Omega))$ be a measurable space, let $\mathfrak{P}(\Omega)$ be the convex set of all probability measures on Ω and let $\mathfrak{O}(\Omega)$ be the set of all real random variables with the natural relation of functional subordination. Let μ_P^X defined by

$$\mu_P^X(B) := P\left(X^{-1}(B)\right), \quad B \in \mathcal{B}(\mathbb{R})$$

be the probability distribution of the random variable $X \in \mathfrak{O}(\Omega)$ with respect to the probability measure $P \in \mathfrak{P}(\Omega)$. The pair $(\mathfrak{P}(\Omega), \mathfrak{O}(\Omega))$ forms a separable statistical model, which we call the *Kolmogorov model*. In this model all observables are compatible and the center is all of $\mathfrak{O}(\Omega)$.

The statistical model for a N-level quantum system was described in Sect. 0.1. If observable X has spectral decomposition (0.2), then the observable $f \circ X$ is naturally defined as

$$f \circ X = \sum_{j=1}^{n} f(x_j) E_j. \tag{0.6}$$

Quantum observables $X_1, \ldots, X_m$ are compatible if and only if the corresponding matrices *commute*, that is

$$X_i X_j = X_j X_i \text{ for } i, j = 1, \ldots, m.$$

The center of this model is trivial: it consists of the matrices which are multiples of the identity; that is, only of the constant observables.

Existence of incompatible observables is a manifestation of the quantum *principle of complementarity*. Physical measurements on microobjects are implemented by macroscopic experimental devices, each of which requires a complex and specific organization of the environment in space and time. Different ways of such organization, corresponding to measurements of different observables, may be mutually exclusive (despite the fact that they relate to the identically prepared microobject); that is, complementary. Complementarity is the first fundamental distinction between quantum and classical statistical models.

One of the most controversial problems of quantum theory is that of "hidden variables"; that is, the question of whether it is possible in principle to describe quantum statistics in terms of a classical probability space. The first attempt to prove the impossibility of introducing hidden variables was made by J. von Neumann in [175] and for some time his arguments were held to be decisive. In 1966 J. Bell, by analyzing Einstein-Podolski-Rosen paradox, had shown the incompleteness of von Neumann's "proof" and distinguished another fundamental property of the quantum-mechanical description which may be called *nonseparability*. Mathematically it is related to the superposition principle and to the fact that composite quantum systems are described by tensor products rather than by the Cartesian products of classical probability theory.

Complementarity and nonseparability are the fundamental properties of quantum statistics underlying non-existence of physically acceptable hidden variables theories (see Sect. 1.4 in Chap. 1).

0.4 Randomization in Classical and Quantum Statistics

Let us consider a classical random variable with a finite set of values $\{x_j\}$. Such a random variable can be represented in the form (0.2), namely

$$X = \sum_{j=1}^{n} x_j E_j \tag{0.7}$$

where $E_j := 1_{B_j}$ is the indicator of the set $B_j := \{\omega \in \Omega : X(\omega) = x_j\}$. The sets B_j $(j = 1, \ldots, n)$ form a partition of Ω. At each point $\omega \in \Omega$ observable (0.7) assumes, with probability 1, one of the values x_j.

In mathematical statistics, in particular, in statistical decision theory, it is useful to consider "randomized" variables which are determined by probabilities $M_j(\omega) \in [0,1]$ of taking the values x_j for the arbitrary elementary event ω. The collection of functions $M := \{M_j : j = 1,\ldots,n\}$ is characterized by the properties

$$M_j(\omega) \geq 0, \quad \sum_{j=1}^{n} M_j(\omega) = 1 \text{ for all } \omega \in \Omega \qquad (0.8)$$

and describes an inexact measurement of the random variable X; that is, a measurement subject to random errors. In particular, the measurement is exact (without error) if $M_j(\omega) = E_j(\omega)$. The probability distribution of this measurement with respect to the probability measure P is given by the formula

$$\mu_P^M(x_j) = \int_\Omega P(d\omega) M_j(\omega) \qquad (0.9)$$

In this way arises another classical statistical model, the *Wald model*, after the originator of statistical decision theory.

It is natural to introduce quantum analogue of the Wald model, in which an observable with a finite set of values is described by a finite *resolution of the identity*; that is, a collection of matrices (operators) $M = \{M_j : j = 1,\ldots,n\}$ in the Hilbert space $\mathcal{H}$ of the system, satisfying the conditions

$$M_j \geq 0, \quad \sum_{j=1}^{n} M_j = 1. \qquad (0.10)$$

In analogy with (0.9) the probability of the j-th outcome in the state described by density operator S is

$$\mu_S^M(x_j) = \operatorname{Tr} SM_j. \qquad (0.11)$$

It is possible to generalize these definitions to observables with arbitrary outcomes. In this way one obtains the *generalized statistical model of quantum mechanics* (see Sect. 2.1 of Chap. 2).

Non-orthogonal resolutions of the identity entered quantum theory in the 70-es through the study of quantum axiomatics [161], [71], the repeatability problem in quantum measurement [56], the quantum statistical decision theory [100] and elsewhere.

0.5 Convex Geometry and Fundamental Limits for Quantum Measurements

In the 1940-50s D. Gabor and L. Brillouin suggested that the quantum-mechanical nature of a communication channel imposes specific fundamental restrictions to the rate of transmission and accuracy of reproduction of information. This problem became important in the 60s with the advent of quantum communication channels – information transmission systems using coherent laser radiation. In the 70s a consistent quantum statistical decision theory was created, which gave the framework for investigation of fundamental limits of accuracy and information content of physical measurement (see [94], [105]). In this theory, the measurement statistics is described by resolutions of the identity in the Hilbert space of the system and the problems of determining the extremum of a functional, such as uncertainty or Shannon information, is solved in a class of quantum measurements subjected to some additional restrictions, such as unbiasedness, covariance etc. (see Chap. 2).On this way the following property of *information openness* was discovered: a measurement over an extension of an observed quantum system, including an independent auxiliary quantum system, may give more information about the state of observed system than any direct measurement of the observed system.

This property has no classical analog - introduction of independent classical systems means additional noise in observations and can only reduce the information - and therefore looks paradoxical. Let us give an explanation from the viewpoint of the generalized statistical model of quantum mechanics.

Although the above generalization of the concept of an observable is formally similar to the introduction of randomized variables in classical statistics, in quantum theory nonorthogonal resolutions of the identity have much more significant role than merely a mean for describing inexact measurements. The set of resolutions of the identity (0.10) is convex. This means that in analogy with mixtures of different states one can speak of mixtures of the different measurement statistics'. Physically such mixtures arise when the measuring device has fluctuating parameters. Both from mathematical and physical points of view, the most interesting and important are the extreme points of the convex set of resolutions of the identity, in which the classical randomness is reduced to an unavoidable minimum. It is these resolutions of the identity which describe the statistics of extremally informative and precise measurements optimizing the corresponding functionals. In the classical case, the extreme points of the set (0.8) coincide with the non-randomized procedures $\{E_j : j = 1, \ldots, n\}$, corresponding to the usual random variables. However, in the quantum case, when the number of outcomes $n > 2$, the extreme points of the set (0.10) are no longer exhausted by the orthogonal resolutions of the identity (see Sect. 1.3 of Chap. 2). Therefore to describe extremally precise and informative quantum measurements, nonorthogonal resolutions of the identity become indispensable.

On the other hand, as M.A. Naimark proved in 1940, an arbitrary resolution of the identity can be extended to an orthogonal one in an enlarged Hilbert space. This enables us to interpret a nonorthogonal resolution of the identity as a standard observable in an extension of the initial quantum system, containing an independent auxiliary quantum system. In this way, a measurement over the extension can be more exact and informative than an arbitrary direct quantum measurement. This fact is strictly connected with the quantum property of nonseparability, mentioned in n. 0.3.

In the 90s, stimulated by the progress in quantum communications, cryptography and computing, the whole new field of quantum information emerged (see e. g. [32], [176]). The emphasis in this field is not only on quantum bounds, but mainly on the new constructive possibilities concealed in the quantum statistics, such as efficient quantum communication protocols and algorithms. A central role here is played by *quantum entanglement*, the strongest manifestation of the quantum nonseparability.

0.6 The Correspondence Problem

One of the difficulties in the standard formulation of quantum mechanics lies in the impossibility of associating with certain variables, such as the time, angle, phase, a corresponding selfadjoint operator in the Hilbert space of the system. The reason for this lies in the Stone-von Neumann uniqueness theorem, which imposes rigid constraints on the spectra of canonically conjugate observables. To this circle of problems also are belong the difficulty with localizability (i.e. with the introduction of covariant position observables) for relativistic quantum particles of zero mass.

By considering nonorthogonal resolutions of the identity as observables subject to the covariance conditions arising from the canonical commutation relations, one can to a significant extent avoid these difficulties (see Chap. 2). The general scheme of this approach is to consider the convex set of resolutions of the identity satisfying the covariance condition, and to single out extreme points of this set minimizing the uncertainty functional in some state. In this way one obtains generalized observables of time, phase etc. described by nonorthogonal resolutions of the identity. In spectral theory nonorthogonal resolutions of the identity arise as generalized spectral measures of non-selfadjoint operators. For example, the operator representing the time observable, turns out to be a maximal Hermitian operator which is not selfadjoint. One may say that at the new mathematical level the generalized statistical model of quantum mechanics justifies a "naive" physical understanding of a real observable as a Hermitian, but not necessarily selfadjoint operator.

0.7 Repeated and Continuous Measurements

In classical probability irreversible transformations of states are described by using conditional probabilities. Let a value x_j be obtained as a result of a measurement of the random variable (0.7). Here the classical state, that is, the probability measure $P(d\omega)$ on Ω is transformed according to the formula of conditional probability

$$P(A) \to P(A \mid B_j) = \frac{\int_A P(d\omega)E_j(\omega)}{\int_\Omega P(d\omega)E_j(\omega)} \quad \text{for } A \in \mathcal{B}(\Omega). \tag{0.12}$$

Clearly, if there is no selection according to the result of the measurement, the state Γ does not change:

$$P(A) \to \sum_{j=1}^n P(A \mid B_j)P(B_j) = P(A). \tag{0.13}$$

In quantum statistics, the situation is qualitatively more complicated. The analog of the transformation (0.12) is the well known *von Neumann projection postulate*

$$S \to S_j = \frac{E_j S E_j}{\operatorname{Tr} S E_j}, \tag{0.14}$$

where S is the density operator of the state before the measurement, and S_j is the density operator after the measurement of the observable (0.2), giving the value x_j. The basis for this postulate is the phenomenological repeatability hypothesis assuming extreme accuracy and minimal perturbation caused by the measurement of the observable X. If the outcome of the measurement is not used for a selection, then the state S is transformed according to the formula analogous to (0.13):

$$S \to \sum_{j=1}^n S_j \mu_S^X(x_j) = \sum_{j=1}^n E_j S E_j. \tag{0.15}$$

However, in general $\sum_{j=1}^n E_j S E_j \neq S$; this means, that the change of the state in the course of quantum measurement does not reduce simply to the change of information but also incorporates an unavoidable and irreversible physical influence of the measuring apparatus upon the system being observed. A number of problems is associated with the projection postulate; leaving aside those of a philosophical character, which go beyond the limits of probabilistic interpretation, we shall comment on concrete problems which are successfully resolved within the framework of generalized statistical models of quantum mechanics.

One principal difficulty is the extension of the projection postulate to observables with continuous spectra. To describe the change of a state under arbitrary quantum measurement, the concept of *instrument*, that is a measure with values in the set of operations on quantum states, was introduced [59].

This concept embraces inexact measurements, not satisfying the condition of repeatability and allows to consider state changes under measurements of observables with continuous spectrum. A resolution of the identity is associated with every instrument, and orthogonal resolutions of the identity correspond to instruments satisfying the projection postulate. The concept of instrument gives a possibility for describing the statistics of arbitrary sequence of quantum measurements.

New clarification is provided for the problem of trajectories, coming back to Feynman's formulation of quantum mechanics. A continuous (in time) measurement process for a quantum observable can be represented as the limit of a "series" of n repeated inexact measurements with precision decreasing as $\sqrt{n}$ [16]. The mathematical description of this limit [115] reveals remarkable parallels with the classical scheme of summation of independent random variables, the functional limit theorems in the probability theory and the Lévy-Khinchin representation for processes with independent increments (see Chap. 4). Most important examples are the analog of the Wiener process - continuous measurement of particle position, and the analog of the Poisson process - the counting process in quantum optics.

An outcome of continuous measurement process is a whole trajectory; the state S_t of the system conditioned upon observed trajectory up to time t - the posterior state - satisfies stochastic nonlinear Schrödinger equation. This equation explains, in particular, why a wave packet does not spread in the course of continuous position measurement [29] (see Chap. 5).

From a physical viewpoint, the whole new fields of the stochastic trajectories approach in quantum optics [43] and decoherent histories approach in foundations of quantum mechanics [77] are closely related to continuous measurement processes.

0.8 Irreversible Dynamics

The reversible dynamics of an isolated quantum system is described by the equation

$$S \to U_t S U_t^{-1}, \quad -\infty < t < \infty, \tag{0.16}$$

where $\{U_t : -\infty < t < \infty\}$ is a continuous group of unitary operators. If the system is *open*, i.e. interacts with its environment, then its evolution is in general irreversible. An example of such an irreversible state change, due to interaction with measuring apparatus, is given by (0.15). The most general form of *dynamical map* describing the evolution of open system and embracing both (0.16) and (0.15) is

$$S \to \sum_j V_j S V_j^*, \tag{0.17}$$

where $\sum_j V_j^* V_j \leq I$. Among all affine transformations of the convex set of states, the maps (0.17) are distinguished by the specifically noncommutative property of *complete positivity*, arising and playing an important part in the modern theory of operator algebras.

A continuous Markov evolution of an open system is described by a *dynamical semigroup*; that is, a semigroup of dynamical maps, satisfying certain continuity conditions (see [56]). Dynamical semigroups, which are the noncommutative analogue of Markov semigroups in probability theory, and their generators describing quantum Markovian master equations, are discussed in Chap. 3.

0.9 Quantum Stochastic Processes

One of the stimuli for the emergence of the theory of quantum stochastic processes was the problem of the *dilation* of a dynamical semigroup to the reversible dynamics of a larger system, consisting of the open system and its environment. By the existence of such a dilation the concept of the dynamical semigroup is rendered compatible with the basic dynamical principle of quantum mechanics given by (0.16).

In probability theory a similar dilation of a Markov semigroup to a group of time shifts in the space of trajectories of a Markov stochastic process is effected by the well-known *Kolmogorov-Daniel construction*. A concept of quantum stochastic process which plays an important role in the problem of the dilation of a dynamical semigroup, was formulated in [2]. In the 80s the theory of quantum stochastic processes turned into a vast field of research (see in particular [194], [195], [196], [197]).

The analytical apparatus of *quantum stochastic calculus*, which in particular permits the construction of nontrivial classes of quantum stochastic processes and concrete dilations of dynamical semigroups, was proposed [138]. Quantum stochastic calculus arises at the intersection of two concepts, namely, the time filtration in the sense of the theory of stochastic processes and the second quantization in the Fock space. It is the structure of continuous tensor product which underlies the connection between infinite divisibility, processes with independent increments and the Fock space. Because of this, the Fock space turns out to be a carrier of the "quantum noise" processes, which give a universal model for environment of an open Markovian quantum system. Quantum stochastic calculus is also interesting from the point of view of the classical theory of stochastic processes. It forms a bridge between Ito calculus and second quantization, reveals an unexpected connection between continuous and jump processes, and provides a fresh insight into the concept of stochastic integral [185], [170]. Finally, on this foundation potentially important applications develop, relating to filtering theory for quantum stochastic processes and stochastic trajectories approach in quantum optics (see Chap. 5).

1. The Standard Statistical Model of Quantum Mechanics

1.1 Basic Concepts

1.1.1 Operators in Hilbert Space

Several excellent books have been devoted to the theory of operators in Hilbert space, to a significant extent stimulated by problems of quantum mechanics (see, in particular, [7], [199]). Here we only recall certain facts and fix our notations.

In what follows $\mathcal{H}$ denotes a separable complex Hilbert space. We use the Dirac notation for the inner product $\langle\varphi|\psi\rangle$, which is linear in ψ and conjugate linear in φ. Sometimes vectors of $\mathcal{H}$ will be denoted as $|\phi\rangle$, and the corresponding (by the Riesz theorem) continuous linear functionals on $\mathcal{H}$ by $\langle\psi|$. In the finite dimensional case these are the column and the row vectors, correspondingly. The symbol $|\psi\rangle\langle\varphi|$ then denotes the rank one operator which acts on a vector $\chi \in \mathcal{H}$ as

$$|\psi\rangle\langle\varphi|\chi = \psi\langle\varphi|\chi\rangle.$$

In particular, if $\langle\psi|\psi\rangle = 1$ then $|\psi\rangle\langle\psi|$ is the projector onto the subspace spanned by the vector $\psi \in \mathcal{H}$. The linear span of the set of operators of the form $|\psi\rangle\langle\varphi|$ is the set of operators of finite rank in $\mathcal{H}$.

The Dirac notation will be used also for densely defined linear ($\langle|$) and antilinear ($|\rangle$) forms on $\mathcal{H}$, which are not given by any vector in $\mathcal{H}$. Let, for example, $\mathcal{H} = L^2(\mathbb{R})$ be the space of square-integrable functions $\psi(x)$ on the real line $\mathbb{R}$. For every $x \in \mathbb{R}$ the relation

$$\langle x|\psi\rangle = \psi(x)$$

defines a linear form $\langle x|$ on the dense subspace $C_0(\mathbb{R})$ of continuous functions with a compact support. This form is unbounded and not representable by any vector in $\mathcal{H}$. Another useful example, which is just the Fourier transform of the previous one is

$$\langle p|\psi\rangle = \frac{1}{\sqrt{2\pi\hbar}} \int_{-\infty}^{\infty} e^{-ipx/\hbar} \psi(x)dx,$$

defined on the subspace of integrable functions.

If X is a bounded operator in $\mathcal{H}$, then X^* denotes its adjoint, defined by

$$\langle \varphi | X^* \psi \rangle = \langle X\varphi | \psi \rangle \text{ for } \varphi, \psi \in \mathcal{H}.$$

A bounded operator $X \in \mathfrak{B}(\mathcal{H})$ is called *Hermitian* if $X = X^*$. An *isometric* operator is an operator U such that $U^*U = I$, where I is the identity operator; if moreover $UU^* = I$, then U is called *unitary*. An (orthogonal) *projection* is a Hermitian operator E such that $E^2 = E$.

The Hermitian operator X is called *positive* $(X \geq 0)$, if $\langle \psi | X\psi \rangle \geq 0$ for all $\psi \in \mathcal{H}$. A positive operator has a unique positive square root, i.e. for any positive operator B there exists one and only one positive operator A such that $A^2 = B$.

For a bounded positive operator T the *trace* is defined by

$$\operatorname{Tr} T := \sum_{i=1}^{\infty} \langle e_i | Te_i \rangle \leq \infty, \tag{1.1}$$

where $\{e_i\}$ is an arbitrary orthonormal Hilbert base. The operator T belongs to the *trace class*, if it is a linear combination of positive operators with finite trace. For such an operator the trace is well defined as the sum of an absolutely convergent series of the form (1.1). The set $\mathfrak{T}(\mathcal{H})$ of all trace class operators is a Banach space with respect to the *trace norm* $\|T\|_1 := \operatorname{Tr} \sqrt{T^*T}$ and the set of operators of finite rank is dense in $\mathfrak{T}(\mathcal{H})$.

The set $\mathfrak{T}(\mathcal{H})$ forms a two-sided ideal in the algebra $\mathfrak{B}(\mathcal{H})$, that is, it is stable under both left and right multiplications by a bounded operator. The dual of the Banach space $\mathfrak{T}(\mathcal{H})$ is isomorphic to $\mathfrak{B}(\mathcal{H})$, the algebra of all bounded operators in $\mathcal{H}$, with the duality given by the bilinear form

$$\langle T, X \rangle = \operatorname{Tr} TX \text{ for } T \in \mathfrak{T}(\mathcal{H}), X \in \mathfrak{B}(\mathcal{H}). \tag{1.2}$$

The subscript h will be applied to sets of operators to denote the corresponding subset of Hermitian operators. For example, $\mathfrak{B}_h(\mathcal{H})$ is the real Banach space of bounded Hermitian operators in $\mathcal{H}$. Then $\mathfrak{T}_h(\mathcal{H})^* = \mathfrak{B}_h(\mathcal{H})$, with the duality given by (1.2) as before.

Along with the convergence in the operator norm in $\mathfrak{B}_h(\mathcal{H})$ some weaker concepts of convergence are often used. One says, the sequence $\{X_n\}$ *converges to X*

- *strongly* if $\lim_{n \to \infty} \|X_n\psi - X\psi\| = 0$ for all $\psi \in \mathcal{H}$,
- *weakly* if $\lim_{n \to \infty} \langle \varphi | X_n\psi \rangle = \langle \varphi | X\psi \rangle$ for all $\varphi, \psi \in \mathcal{H}$ and
- **-weakly* if $\lim_{n \to \infty} \operatorname{Tr} TX_n = \operatorname{Tr} TX$ for all $T \in \mathfrak{T}(\mathcal{H})$.

If a sequence of operators $\{X_n\}$ is bounded with respect to the norm and $X_n \leq X_{n+1}$ for all n, then X_n converges strongly, weakly and *-weakly to a bounded operator X (denoted $X_n \uparrow X$).

1.1.2 Density Operators

These are the positive operators S of unit trace

$$S \geq 0, \quad \text{Tr}\, S = 1.$$

The set of density operators $\mathfrak{S}(\mathcal{H})$ is a convex subset of the real linear space $\mathfrak{T}_h(\mathcal{H})$ of the Hermitian trace class operators. Furthermore it is the base of the cone of positive elements which generates $\mathfrak{T}_h(\mathcal{H})$. A point S of a convex set $\mathfrak{S}$ called *extreme* if from $S = pS_1 + (1-p)S_2$, where $S_1, S_2 \in \mathfrak{S}$ and $0 < p < 1$, it follows that $S_1 = S_2 = S$. The extreme points of the sets $\mathfrak{S}(\mathcal{H})$ are the one-dimensional projectors

$$S_\psi = |\psi\rangle\langle\psi|, \tag{1.3}$$

where $\psi \in \mathcal{H}$ and $\langle\psi|\psi\rangle = 1$. Any density operator can be represented as a convex combination

$$S = \sum_{j=1}^{\infty} p_j |\psi_j\rangle\langle\psi_j|,$$

where $\langle\psi_j|\psi_j\rangle = 1$, $p_j \geq 0$ and $\sum_{j=1}^{\infty} p_j = 1$. One such representation is given by the spectral decomposition of the operator S, when the ψ_j are its eigenvectors and the p_j are the corresponding eigenvalues.

The *entropy* of the density operator S is defined as

$$H(S) = -\text{Tr}\, S \log S = -\sum_j \lambda_j \log \lambda_j,$$

where λ_j are the eigenvalues of S, with the convention $0 \log 0 = 0$ (usually $\log$ denotes the binary logarithm). The entropy is a nonnegative concave function on $\mathfrak{S}(\mathcal{H})$, taking its minimal value 0 on the extreme points of $\mathfrak{S}(\mathcal{H})$. If $d = \dim \mathcal{H} < \infty$, then the maximal value of the entropy is $\log d$, and it is achieved on the density operator $S = d^{-1}I$.

Consider the set $\mathfrak{E}(\mathcal{H})$ of projectors in $\mathcal{H}$, which is isomorphic to the quantum logic of events (closed linear subspaces of $\mathcal{H}$). A *probability measure* on $\mathfrak{E}(\mathcal{H})$ is a real function μ with the properties

1. $0 \leq \mu(E) \leq 1$ for all $E \in \mathfrak{E}(\mathcal{H})$,
2. if $\{E_j\} \subset \mathfrak{E}(\mathcal{H})$ and $E_j E_k = 0$ whenever $j \neq k$ and $\sum_j E_j = I$ then $\sum_j \mu(E_j) = I$.

In respond to a question of Mackey, Gleason proved the following theorem (see [162], [185])

Theorem 1.1.1. *Let* $\dim \mathcal{H} \geq 3$. *Then any probability measure* μ *on* $\mathfrak{E}(\mathcal{H})$ *has the form*

$$\mu(E) = \text{Tr } SE \text{ for all } E \in \mathfrak{E}(\mathcal{H}), \tag{1.4}$$

where S is a uniquely defined density operator.

The case $\dim \mathcal{H} = 2$ is singular - for it one easily shows that there are measures not representable in the form (1.4), however they did not find application in quantum theory. The proof of Gleason's theorem is quite nontrivial and generated a whole field in noncommutative measure theory, dealing with possible generalizations and simplifications of this theorem. (See e. g. Kruchinsky's review in [193]).

1.1.3 The Spectral Measure

Let $\mathcal{X}$ be a set equipped with a σ-algebra of measurable subsets $\mathfrak{B}(\mathcal{X})$. An *orthogonal resolution of the identity* in $\mathcal{H}$ is a projector-valued measure on $\mathfrak{B}(\mathcal{X})$, that is a function $E : \mathfrak{B}(\mathcal{X}) \to \mathfrak{E}(\mathcal{H})$ satisfying the conditions

1. if $B_1, B_2 \in \mathfrak{B}(\mathcal{X})$ and $B_1 \cap B_2 = \emptyset$, then $E(B_1)E(B_2) = 0$,
2. if $\{B_j\}$ is a finite or countable partition of $\mathcal{X}$ into pairwise nonintersecting measurable subsets, then $\sum_j E(B_j) = I$, where the series converges strongly.

Let X be an operator with dense domain $\mathcal{D}(X) \subset \mathcal{X}$. We denote by $\mathcal{D}(X^*)$ the set of vectors φ for which there exists $\chi \in \mathcal{H}$ such that

$$\langle \varphi | X \psi \rangle = \langle \chi | \psi \rangle \text{ for all } \psi \in \mathcal{D}(X).$$

We define the operator X^* with domain $\mathcal{D}(X^*)$ by $X^* \varphi = \chi$. The operator X is said to be *Hermitian* (symmetric) if $X \subseteq X^*$ ($\mathcal{D}(X) \subset \mathcal{D}(X)^*$ and $X = X^*$ on $\mathcal{D}(X)$), and *selfadjoint* if $X = X^*$.

The *spectral theorem* (von Neumann, Stone, Riesz, 1929-1932) establishes a one-to-one correspondence between orthogonal resolutions of identity E on the σ-algebra $\mathcal{B}(\mathbb{R})$ of Borel subsets of the real line $\mathbb{R}$ and selfadjoint operators in $\mathcal{H}$ according to the formula

$$X = \int\limits_{-\infty}^{\infty} x E(dx), \tag{1.5}$$

where the integral is defined in appropriate sense (see [199]). The resolution of the identity E is said to be the *spectral measure* of the operator X. For an arbitrary Borel function f one defines the selfadjoint operator

$$f \circ X := \int\limits_{-\infty}^{\infty} f(x)E(dx).$$

The spectral measure F of $f \circ X$ is related to the spectral measure of the operator X by

$$F(B) = E\big(f^{-1}(B)\big) \text{ for } B \in \mathcal{B}(\mathbb{R}).$$

We denote by $\mathfrak{O}(\mathcal{H})$ the set of all selfadjoint operators in $\mathcal{H}$.

1.1.4 The Statistical Postulate

With every quantum mechanical system there is associated a separable complex Hilbert space $\mathcal{H}$. The *states* of the system are described by the density operators in $\mathcal{H}$ (elements of $\mathfrak{S}(\mathcal{H})$). Extreme points of $\mathfrak{S}(\mathcal{H})$ are called *pure* states. A *real observable* is an arbitrary selfadjoint operator in $\mathcal{H}$ (element of $\mathfrak{O}(\mathcal{H})$). The probability distribution of the observable X in the state S is defined by the Born-von Neumann formula

$$\mu_S^X(B) = \operatorname{Tr} SE(B) \quad \text{for } B \in \mathcal{B}(\mathbb{R}), \tag{1.6}$$

where E is the spectral measure of X. We call the separable statistical model $(\mathfrak{S}(\mathcal{H}), \mathfrak{O}(\mathcal{H}))$ thus defined the *standard statistical model of quantum mechanics*.

From (1.5) and (1.6) it follows that the *mean value* of the observable X in the state S

$$\mathbf{E}_S(X) = \int\limits_{-\infty}^{\infty} x \mu_S^X(dx)$$

is given by

$$\mathbf{E}_S(X) = \operatorname{Tr} SX \tag{1.7}$$

(at least for bounded observables). The mean value of the observable in a pure state is given by the matrix element

$$\mathbf{E}_{S_\psi}(X) = \langle \psi | X\psi \rangle.$$

Abusing terminology one sometimes calls an arbitrary element $X \in \mathfrak{B}(\mathcal{H})$ a bounded observable. The relation (1.7) defines a positive linear normalized

($E_S(I) = 1$) functional on the algebra $\mathfrak{B}(\mathcal{H})$; that is, a *state* in the sense of the theory of algebras (the preceding argument explains the origin of this mathematical terminology). A state on $\mathfrak{B}(\mathcal{H})$ defined by a density operator by (1.7) is *normal*, meaning that if $X_n \uparrow X$, then $E_S(X_n) \to E_S(X)$.

A *von Neumann algebra* is an algebra of bounded operators in $\mathcal{H}$ which contains the unit operator and is closed under involution and under transition to limits in the strong (or weak) operator topology. For an arbitrary von Neumann algebra $\mathfrak{B}$, just as for $\mathfrak{B}(\mathcal{H})$, there is an associated statistical model in which the states are normal states on $\mathfrak{B}$, while the observables are selfadjoint operators affiliated with $\mathfrak{B}$. Such models occupy an intermediate place between quantum and classical (when $\mathfrak{B}$ is commutative), and play an important role in the theories of quantum systems of infinitely many degrees of freedom - the quantum field theory and statistical mechanics (see e. g. [36], [40], [66]).

1.1.5 Compatible Observables

The *commutator* of bounded operators X and Y is the operator

$$[X, Y] := XY - YX.$$

Operators X, Y *commute* if $[X, Y] = 0$. Selfadjoint operators X and Y are called commuting if their spectral measures commute.

The following statements are equivalent:

1. The observables $X_1, \ldots, X_n$ are *compatible*, that is, there exists an observable X and Borel functions $f_1, \ldots, f_n$ such that $X_j = f_j \circ X$ for $j = 1, \ldots, n$.
2. The operators $X_1, \ldots, X_n$ commute.

Indeed, if $E_1, \ldots, E_n$ are the spectral measures of compatible observables $X_1, \ldots, X_n$, then there is a unique orthogonal resolution of the identity E on $\mathcal{B}(\mathbb{R}^n)$ for which

$$E(B_1 \times \cdots \times B_n) = E_1(B_1) \cdots E_n(B_n) \text{ for } \{B_j : j = 1, \ldots, n\} \subset \mathcal{B}(\mathbb{R}),$$

called the joint spectral measure of the operators $X_1, \ldots, X_n$.

The existence of incompatible observables is a manifestation of the quantum complementarity. A quantitative expression for it is given by the *uncertainty relation*. For observables X, Y having finite second moments with respect to the state S (the n-th moment of a observable X with respect to the state S is defined by $\int x^n \mu_S^X(dx)$), the following bilinear forms are well defined:

$$\langle X, Y \rangle_S := \Re(\operatorname{Tr} YSX), \quad [X, Y]_S := 2\Im(\operatorname{Tr} YSX)$$

(see [105] Chap. 2). Let $\mathcal{X} := \{X_1, \ldots, X_n\}$ be an arbitrary collection of observables with finite second moments. Let us introduce the real matrices

$$\mathbf{D}_S(\mathcal{X}) := \Big(\langle X_i - I\mathbf{E}_S(X_i), X_j - I\mathbf{E}_S(X_j) \rangle_S \Big)_{i,j=1,\ldots,n}$$

$$\mathbf{C}_S(\mathcal{X}) := \Big([X_i, X_j]_S \Big)_{i,j=1,\ldots,n}.$$

From positive definiteness of the sesquilinear forms $(X, Y) \to \operatorname{Tr} Y^* SX$ and $(X, Y) \to \operatorname{Tr} XSY^*$ follows the inequality

$$\mathbf{D}_S(\mathcal{X}) \geq \pm \frac{i}{2} \mathbf{C}_S(\mathcal{X}), \tag{1.8}$$

where the left and the right hand side are considered as complex Hermitian matrices[1]. For two observables $X = X_1$ and $Y = X_2$ the inequality (1.8) is equivalent to the Schrödinger-Robertson uncertainty relation

$$\mathbf{D}_S(X)\mathbf{D}_S(Y) \geq \langle X - I\mathbf{E}_S(X), Y - I\mathbf{E}_S(Y) \rangle^2 + \frac{1}{4}[X, Y]_S^2, \tag{1.9}$$

where

$$\mathbf{D}_S(X) = \int\limits_{-\infty}^{\infty} \left(x - \mathbf{E}_S(X) \right)^2 \mu_S^X(dx)$$

is the *variance* of the observable X in the state S. If X, Y are compatible observables then the quantity

$$\langle X - I\mathbf{E}_S(X), Y - I\mathbf{E}_S(Y) \rangle_S = \int\limits_{-\infty}^{\infty} \int\limits_{-\infty}^{\infty} \left(x - \mathbf{E}_S(X) \right) \left(y - \mathbf{E}_S(Y) \right) \mu_S^{X,Y}(dxdy)$$

represents the *covariance* of X, Y in the state S; in this case $[X, Y]_S = 0$ and (1.9) reduces to the Cauchy-Schwarz inequality for covariances of random variables. For arbitrary bounded Hermitian operators X, Y

---

[1] This inequality is due to Robertson; it was rediscovered later by many authors (see [64]).

$$\langle X, Y \rangle_S = \text{Tr } S X \circ Y, \tag{1.10}$$

where

$$X \circ Y := \frac{1}{2}(XY + YX)$$

is the *Jordan (symmetrized) product* of the operators X, Y. Quantities of the form (1.10) are called *correlations* in quantum statistical mechanics, although if the observables X, Y are not compatible, they are not associated in any simple way with measurement statistics of X and Y.

A detailed review of various generalizations of the uncertainty relation can be found in [64].

1.1.6 The Simplest Quantum System

In physics, a finite-dimensional Hilbert space usually describes internal degrees of freedom (spin) of a quantum system. The case $\dim \mathcal{H} = 2$ corresponds to the minimal nonzero spin $\frac{1}{2}$. Consider two-dimensional Hilbert space $\mathcal{H}$ with the canonical basis

$$|\uparrow\rangle = \begin{pmatrix} 1 \\ 0 \end{pmatrix}, \quad |\downarrow\rangle = \begin{pmatrix} 0 \\ 1 \end{pmatrix}. \tag{1.11}$$

An important basis in $\mathfrak{O}(\mathcal{H})$ is given by the matrices

$$I := \begin{pmatrix} 1 & 0 \\ 0 & 1 \end{pmatrix}, \quad \sigma_1 := \begin{pmatrix} 0 & 1 \\ 1 & 0 \end{pmatrix}, \quad \sigma_2 := \begin{pmatrix} 0 & -i \\ i & 0 \end{pmatrix}, \quad \sigma_3 := \begin{pmatrix} 1 & 0 \\ 0 & -1 \end{pmatrix}.$$

The matrices σ_i called the *Pauli matrices* arise from a unitary representation of euclidean rotations in $\mathcal{H}$ and express geometry of the particle spin. In quantum computation this simplest system describes an elementary memory cell of a quantum computer - the *qubit* [212], [32]. Here the action of Pauli matrices describes elementary errors which can happen in this cell: the spin flip σ_1, the phase error σ_3, and their combination $\sigma_2 = i\sigma_1\sigma_3$.

If $X(a) := \sum_{i=1}^{3} a_i \sigma_i$, where $a = (a_1, a_2, a_3) \in \mathbb{R}^3$, is a traceless matrix, we have

$$X(a)X(b) = (a \cdot b)I + iX(a \times b), \tag{1.12}$$

where $a \cdot b$ is the scalar product, and $a \times b$ the vector product of the vectors a, b. Hence

$$\operatorname{Tr} X(a)X(b) = 2a \cdot b. \tag{1.13}$$

Each density matrix can be expressed uniquely in the form

$$S(a) = \frac{1}{2}(I + X(a)), \tag{1.14}$$

where $|a| \leq 1$. Thus as a convex set $\mathfrak{S}(\mathcal{H})$ is isomorphic to the unit ball in $\mathbb{R}^3$, moreover the pure states correspond to points of the sphere $|a| = 1$. In this case

$$S(a) = |\psi(a)\rangle\langle\psi(a)|, \tag{1.15}$$

where

$$\psi(a) = \begin{bmatrix} \cos(\beta/2)e^{(-i\alpha/2)} \\ \sin(\beta/2)e^{(i\alpha/2)} \end{bmatrix}$$

is the unit vector of the state, and $\cos\beta = a_3, \sin\beta e^{i\alpha} = a_1 + ia_2$. The parameters α, β are the Euler angles of the unit vector $a \in \mathbb{R}^3$, describing the direction of the particle spin. Pure states with $|a| = 1$ are "completely polarized" states with the definite direction of the spin, while their mixtures with $|a| \leq 1$ are "partially polarized". To $a = 0$ corresponds the "chaotic state" $S(0) = \frac{1}{2}I$. A natural particle source usually provides chaotic state, while a completely polarized state is prepared by "Stern-Gerlach filter" involving application of an inhomogeneous magnetic field with gradient in the direction a [187].

Since from (1.14) using (1.13) $\langle\psi(a)|\psi(-a)\rangle = \operatorname{Tr} S(a)S(-a) = 0$, the spectral decomposition of the observable $X(a)$ is

$$X(a) = |\psi(a)\rangle\langle\psi(a)| - |\psi(-a)\rangle\langle\psi(-a)|, \qquad (|a| = 1).$$

Thus the observable $X(a)$ assumes the values ± 1, with the probabilities

$$\operatorname{Tr} S(b)S(\pm a) = \frac{1}{2}(1 \pm a \cdot b)$$

in the state $S(b)$ ($|b| < 1$).

The observable $\frac{\hbar}{2}X(a)$ (where $\hbar$ is the Dirac constant) describes the spin component in the direction a. From (1.12) it follows that the observables $X(a), X(b)$ are compatible if and only if a and b are collinear. The Stern-Gerlach devices measuring $X(a), X(b)$ require application of inhomogeneous magnetic fields with corresponding gradients, and hence are complementary to each other.

1.2 Symmetries, Kinematics, Dynamics

1.2.1 Groups of Symmetries

Consider a separable statistical model $(\mathfrak{S}, \mathfrak{O})$. Let there be given a pair of bijective (one-to-one onto) mappings $\Psi : \mathfrak{S} \to \mathfrak{S}$ and $\Phi : \mathfrak{O} \to \mathfrak{O}$ such that

$$\mu_{\Psi(S)}^{\Phi(X)} = \mu_S^X$$

for all $S \in \mathfrak{S}$ and $X \in \mathfrak{O}$. From this it follows that Ψ is an affine map, that is

$$\Psi\left(\sum_{i=1}^{n} p_i S_i\right) = \sum_{i=1}^{n} p_i \Psi(S_i),$$

for $p_i \geq 0$, $\sum_{i=1}^{n} p_i = 1$. We shall call the map Ψ a *symmetry* in the state space, and the map Φ the corresponding symmetry in the space of observables.

Theorem 1.2.1 (Wigner). *Every symmetry of the space of quantum states $\mathfrak{S}(\mathcal{H})$ has the form*

$$\Psi(S) = USU^*, \tag{1.16}$$

where U is a unitary or antiunitary operator in $\mathcal{H}$.

For mean values of observables we have

$$\mathbf{E}_{\Psi(S)}(X) = \operatorname{Tr} \Psi(S) X = \operatorname{Tr} S \Psi^*(X) = \mathbf{E}_S(\Psi^*(X)),$$

where

$$\Psi^*(X) = U^* X U. \tag{1.17}$$

From the statistical viewpoint the transformation (1.16) of states is equivalent to the transformation (1.17) of observables. In the first case one speaks of the *Schrödinger picture* and in the second of the *Heisenberg picture*.

Let G be a group and let $g \to \Psi_g$ a map from G into the group of symmetries of $\mathfrak{S}(\mathcal{H})$ such that

$$\Psi_{g_1 \cdot g_2} = \Psi_{g_1} \cdot \Psi_{g_2} \quad \text{for all } g_1, g_2 \in G.$$

If G is a connected topological group and the map $g \rightarrow \Psi_g$ is continuous, then Ψ_g can be represented as

$$\Psi_g(S) = U_g S U_g^*,$$

where $g \rightarrow U_g$ is a projective unitary representation of the group G in the space $\mathcal{H}$; that is, all U_g are unitary operators and satisfy the equation

$$U_{g_1} U_{g_2} = \omega(g_1, g_2) U_{g_1 g_2}, \qquad (1.18)$$

where $(g_1, g_2) \rightarrow \omega(g_1, g_2)$ - the multiplier of the representation - is a complex function obeying certain algebraic relations (see, for example [222]).

The second alternative in the Wigner theorem can be rephrased in the following way: let us fix some orthonormal base in $\mathcal{H}$ and let Λ be the antiunitary operator of complex conjugation in this base. Then $\Lambda^2 = I$ and $\tilde{U} = \Lambda U$ is unitary, therefore

$$\Psi(S) = \tilde{U}^* \bar{S} \tilde{U} = \tilde{U}^* S' \tilde{U},$$

where $\bar{S}$ is complex conjugation, and S' is transposition of the operator's matrix in the base. Thus the source of antiunitarity is complex conjugation, which is associated in physics with the time inversion [36].

1.2.2 One-Parameter Groups

In the case $G = \mathbb{R}$ it is always possible to choose $\omega(g_1, g_2) \equiv 1$, so that a one-parameter group of symmetries always gives rise to a unitary representation of $\mathbb{R}$ in $\mathcal{H}$ [222].

Theorem 1.2.2 (Stone). *Let $t \rightarrow U_t$, $t \in \mathbb{R}$ be a strongly continuous group of unitary operators in $\mathcal{H}$ so that*

$$U_{t_1} U_{t_2} = U_{t_1 + t_2} \text{ for all } t_1, t_2 \in \mathbb{R}.$$

Then there exists a selfadjoint operator A in $\mathcal{H}$ such that $U_t = e^{itA}$ for all $t \in \mathbb{R}$. Conversely, for an arbitrary selfadjoint operator A the family $\{e^{itA} : t \in \mathbb{R}\}$ forms a strongly continuous one-parameter group.

Consider, for example, the group of spatial shifts along the given coordinate axis. Let S_0 be the state prepared by a certain device, then the new state prepared by the same device shifted through a distance x along the axis is

$$e^{-iPx/\hbar} S_0 e^{iPx/\hbar}, \qquad (1.19)$$

where P is the selfadjoint operator of *momentum* along the axis. Here x represents the parameter of the spatial shift.

On the other hand, let the state S_0 be prepared at a certain time τ. Then the state prepared by the same device at the time $\tau + t$ is

$$e^{iHt/\hbar} S_0 e^{-iHt/\hbar}, \qquad (1.20)$$

where H is the selfadjoint operator called the *Hamiltonian* (representing the energy observable of the system). Since the time shift $\tau \to \tau + t$ in state preparation is equivalent to the time shift $\tau \to \tau - t$ in observation, the time evolution of the state is given by

$$S_0 \to S_t = e^{-iHt/\hbar} S_0 e^{iHt/\hbar},$$

Then the infinitesimal time evolution equation (in the Schrödinger picture) is

$$i\hbar \frac{dS_t}{dt} = [H, S_t]. \qquad (1.21)$$

If $S_0 = |\psi_0\rangle\langle\psi_0|$ is a pure state, then $S_t = |\psi_t\rangle\langle\psi_t|$ for arbitrary t, where $\{\psi_t\}$ is a family of vectors in $\mathcal{H}$, satisfying the Schrödinger equation

$$i\hbar \frac{d\psi_t}{dt} = H\psi_t. \qquad (1.22)$$

In general H is unbounded operator, so that in (1.21) and (1.22) domain considerations are necessary.

In general, to every one-parameter group of symmetries of geometrical or kinematic character there corresponds a selfadjoint operator, which is the generator of transformations of quantum states according to formulas of the type (1.20) and (1.19).

1.2.3 The Weyl Relations

The kinematics of non-relativistic systems is based on the principle of Galilean relativity, according to which the description of an isolated system is the same in all inertial coordinate frames. Let $W_{x,y}$ be the unitary operator describing the transformation of states prepared by the device shifted through a distance x and moving relative to the original frame with velocity $v = y/m$ along a fixed coordinate axis (m is the mass of the particle). Then according to Sect. 1.2.1 $(x, y) \to W_{x,y}$ is a projective representation of the group $G = \mathbb{R}^2$ in the Hilbert space $\mathcal{H}$. If the representation is irreducible, i. e. if there is no closed subspace of $\mathcal{H}$ invariant under all $W_{x,y}$, then one can prove that (1.18) can be chosen in the form

$$W_{x_1,y_1}W_{x_2,y_2} = \exp\left(-\frac{i}{2\hbar}(x_1y_2 - x_2y_1)\right)W_{x_1+x_2,y_1+y_2}. \qquad (1.23)$$

By selecting one-parameter subgroups - the group of spatial shifts $V_x = W_{x,0}$ and the group of Galilean boosts $U_y = W_{0,y}$ - relation (1.23) implies

$$U_yV_x = e^{ixy/\hbar}V_xU_y \qquad \text{for } x, y \in \mathbb{R}, \qquad (1.24)$$

moreover $W_{x,y} = e^{\frac{ixy}{2\hbar}}V_xU_y$. The relation (1.24) is called the *Weyl canonical commutation relation*(CCR) [229].

According to Stone's theorem there exist selfadjoint operators Q and P in $\mathcal{H}$ such that

$$U_y = e^{iyQ/\hbar}, \quad V_x = e^{-ixP/\hbar}.$$

Considering Q as a real observable, we note that (1.24) is equivalent to

$$V_x^* E(B + x)V_x = E(B), \quad B \in \mathcal{B}(\mathbb{R}) \qquad (1.25)$$

for the spectral measure E of the operator Q. But this is equivalent to the fact that the probability distribution of the observable is transformed covariantly under the spatial shifts (1.19):

$$\mu_{S_x}^Q(B + x) = \mu_{S_0}^Q(B), \quad B \in \mathcal{B}(\mathbb{R}),$$

for an arbitrary state S_0. This allows us to call Q the *position observable* along the chosen axis. A similar argument shows that $\frac{P}{m}$ is the *velocity observable* of the isolated quantum system.

Any pair (V, U) of unitary families, satisfying the relations (1.24) is called a representation of CCR. The following *Schrödinger representation* in $\mathcal{H} = L^2(\mathbb{R})$ is irreducible

$$V_x\psi(\xi) = \psi(\xi - x), \quad U_x\psi(\xi) = e^{iy\xi/\hbar}\psi(\xi).$$

In this representation Q is the operator of multiplication by ξ and P is the operator $\frac{\hbar}{i}\frac{d}{d\xi}$. The operators P, Q have a common dense invariant domain of definition and satisfy on it the *Heisenberg CCR*:

$$[Q, P] = i\hbar I. \tag{1.26}$$

The operators P, Q are called *canonical observables*. The transition in the opposite direction from the Heisenberg relation to the Weyl (exponential) relation involves analytical subtleties, associated with the unboundedness of the operators P and Q, and has given rise to an extensive mathematical literature (see [141]).

Of principal significance for quantum mechanics is the fact that the CCR define the canonical observables P, Q essentially uniquely [162] :

Theorem 1.2.3 (Stone-von Neumann). *Every strongly continuous representation of CCR is the direct sum of irreducible representations, each of which is unitarily equivalent to the Schödinger representation.*

In particular, in any representation of the CCR, as in the Schrödinger representation, the operators P, Q are unbounded and have Lebesgue spectrum extended over the whole real line. Associated with this is the well known difficulty in establishing a correspondence for different canonical pairs in quantum mechanics. The problem is to define canonically conjugate quantum observables, analogous to generalized coordinates and momenta in the Hamiltonian formalism, and is related to the problem of quantization of classical systems. For example, time and energy, like position and momentum, are canonically conjugate in classical mechanics. However the energy observable H has spectrum bounded from below, hence, by the Stone-von Neumann theorem, that there is no selfadjoint operator T representing time, which is related to H by the canonical commutation relation. These difficulties, which arise also for other canonical pairs, can be resolved within the framework of the generalized statistical model of quantum mechanics (see Sect. 2.3 of Chap. 2).

The CCR for systems with arbitrary number of degrees of freedom are formulated in a similar way. Let $\{Z, \Delta\}$ be a *symplectic* space, that is a real linear space with a bilinear skew-symmetric form $(z, z') \rightarrow \Delta(z, z') : Z \times Z \rightarrow \mathbb{R}$. A representation of CCR is a family $z \rightarrow W(z)$ of unitary operators in the Hilbert space $\mathcal{H}$, satisfying the *Weyl-Segal CCR*

$$W(z)W(z') = \exp\left(\frac{i}{2\hbar}\Delta(z, z')\right)W(z + z'). \tag{1.27}$$

If Z is finite-dimensional and the form Δ is non-degenerate, then Z has even dimension $2d$, and there exists a basis $\{b_1, \ldots, b_d, c_1, \ldots, c_d\}$ in which

$$\Delta(z, z') = \sum_{j=1}^{d}(x_j y_j' - x_j' y_j),$$

where $z = \sum_{j=1}^{d} x_j b_j + y_j c_j$ and $z' = \sum_{j=1}^{d} x'_j b_j + y'_j c_j$. Here the operators of a strongly continuous Weyl representation can be written as

$$W(z) = \exp\left(\frac{i}{\hbar} \sum_{j=1}^{d} (x_j P_j + y_j Q_j)\right),$$ (1.28)

where P_j, Q_j are selfadjoint operators satisfying the multidimensional analogue of the Heisenberg CCR (1.26)

$$[Q_j, P_k] = i\delta_{jk}\hbar I, \quad [P_j, P_k] = 0, \quad [Q_j, Q_k] = 0.$$

The Stone-von Neumann uniqueness theorem also holds in this case.

For systems with an infinite number of degrees of freedom the uniqueness is violated and there exists a continuum of inequivalent representations, which is the cause of "infrared divergences" in quantum field theory (see e. g. [36], [206], [66]). This non-uniqueness is closely related to the possible inequivalence of probability (Gaussian) measures in infinite-dimensional spaces and served as one of the initial motivations for the study of this problem, which occupies a substantial place in the classical theory of random processes.

1.2.4 Gaussian States

The inequality (1.9) and the CCR (1.26) lead to the *Heisenberg uncertainty relation*

$$\mathbf{D}_S(P)\mathbf{D}_S(Q) \geq \frac{\hbar^2}{4},$$ (1.29)

from where it follows that there exists no state S in which P and Q simultaneously assume exact values with probability 1. The states for which equality is attained in (1.29), are called the *minimal uncertainty states*. These are the pure states $S_{x,y} = |\psi_{x,y}\rangle\langle\psi_{x,y}|$ defined by the vectors

$$|\psi_{x,y}\rangle = W_{x,y}|\psi_{0,0}\rangle, \quad (x,y) \in \mathbb{R}^2,$$ (1.30)

where $|\psi_{0,0}\rangle$ is the vector of the *ground state*, given by the function

$$\langle\xi|\psi_{0,0}\rangle = (2\pi\sigma^2)^{-\frac{1}{4}} \exp\left(-\frac{\xi^2}{4\sigma^2}\right).$$

in the Schrödinger representation.

The states $S_{x,y}$ are characterized by the three parameters

$$x = \mathbf{E}_{S_{x,v}}(Q)$$
$$y = \mathbf{E}_{S_{x,v}}(P)$$
$$\sigma^2 = \mathbf{D}_{S_{x,v}}(Q) = \hbar^2 \left(4\mathbf{D}_{S_{x,v}}(P)\right)^{-1}.$$

For fixed σ^2 the vectors $\psi_{x,y}$ form a family equivalent to the well known family of coherent states in quantum optics ([82], [146]). For these states

$$\left\langle Q - \mathbf{E}_{S_{x,v}}(Q) \cdot I, P - \mathbf{E}_{S_{x,v}}(P) \cdot I \right\rangle_{S_{x,y}} = 0 \qquad (1.31)$$

A larger class is formed by the pure states for which equality is achieved in the uncertainty relations (1.9) (for P and Q), but (1.31) is not necessarily satisfied. These states have been widely discussed in physics literature under the name of *squeezed states* (see, for example, [191]).

From a mathematical point of view all these states, as well as their thermal mixtures, are contained in the class of states that are a natural quantum analogue of Gaussian distribution in probability theory [99], [6]. Let $\{Z, \Delta\}$ be a finite dimensional symplectic space with nondegenerate skew-symmetric form Δ and let $z \to W(z)$ be an irreducible CCR representation in the Hilbert space $\mathcal{H}$. The *characteristic function* of the density operator $S \in \mathcal{H}$ is defined by

$$\varphi(z) := \operatorname{Tr} SW(z) \quad \text{for } z \in Z, \qquad (1.32)$$

and has several properties similar to those of characteristic functions in probability theory (see e.g. [105], Chap. 5; [227]). In particular, the analog of the condition of positive definiteness has the form

$$\sum_{j,k} \bar{c}_j c_k \varphi(z_j - z_k) \exp\left(\frac{i}{2\hbar}\Delta(z_j, z_k)\right) \geq 0 \qquad (1.33)$$

for all finite sets $\{c_j\} \subset \mathbb{C}$, $\{z_j\} \subset Z$. The state S is *Gaussian* if its characteristic function has the form

$$\varphi(z) = \exp\left(im(z) - \frac{1}{2}\alpha(z, z)\right),$$

where $m(z)$ is a linear and $\alpha(z, z')$ a bilinear form on Z. This form defines a characteristic function if and only if the relation

$$\alpha(z, z)\alpha(z', z') \geq \frac{1}{4}[\Delta(z, z')/\hbar]^2 \qquad \text{for } z, z' \in Z,$$

directly related to (1.33), is satisfied.

In quantum field theory such states describe quasi-free fields and are called quasi-free states. In statistical mechanics they occur as the equilibrium states of Bose-systems with quadratic Hamiltonians (see, for example, [41], [66]).

1.3 Composite Systems

1.3.1 The Tensor Product of Hilbert Spaces

Let $\mathcal{H}_1, \mathcal{H}_2$ be Hilbert spaces with scalar products $\langle\cdot|\cdot\rangle_1, \langle\cdot|\cdot\rangle_2$. Consider the set L of formal linear combinations of the elements $\psi_1 \times \psi_2 \in \mathcal{H}_1 \times \mathcal{H}_2$, and introduce the scalar product in L defining

$$\langle\varphi_1 \times \varphi_2 | \psi_1 \times \psi_2\rangle = \langle\varphi_1|\psi_1\rangle_1 \cdot \langle\varphi_2|\psi_2\rangle_2$$

and extending this to L by linearity. The completion of L (factorized by the null-space of the form) is a Hilbert space called the tensor product of the Hilbert spaces $\mathcal{H}_1, \mathcal{H}_2$ and denoted $\mathcal{H}_1 \otimes \mathcal{H}_2$. The vector in $\mathcal{H}_1 \otimes \mathcal{H}_2$ corresponding to the equivalence class of the element $\psi_1 \times \psi_2 \in \mathcal{H}_1 \times \mathcal{H}_2$ is denoted $\psi_1 \otimes \psi_2$.

For example, if $\mathcal{H}_1 = L^2(\Omega_1, \mu_1)$, $\mathcal{H}_2 = L^2(\Omega_2, \mu_2)$, then the space $\mathcal{H}_1 \otimes \mathcal{H}_2 = L^2(\Omega_1 \times \Omega_2, \mu_1 \times \mu_2)$ consists of all functions $(\omega_1, \omega_2) \to \psi(\omega_1, \omega_2)$, square-integrable with respect to the measure $\mu_1 \times \mu_2$, and the vector $\psi_1 \otimes \psi_2$ is defined by the function $(\omega_1, \omega_2) \to \psi_1(\omega_1)\psi_2(\omega_2)$.

If the $\mathcal{H}_j$ are finite dimensional complex Hilbert spaces then

$$\dim \mathfrak{D}(\mathcal{H}_1 \otimes \mathcal{H}_2) = \dim \mathfrak{D}(\mathcal{H}_1) \dim \mathfrak{D}(\mathcal{H}_2).$$

But if the $\mathcal{H}_j$ are real Hilbert spaces then $=$ becomes $>$, while for quaternion Hilbert spaces (given some reasonable definition of $\mathcal{H}_1 \otimes \mathcal{H}_2$), it becomes $<$. This fact is sometimes regarded as an indirect argument for use of the field of complex numbers in axiomatic quantum mechanics.

The tensor product $\mathcal{H}_1 \otimes \cdots \otimes \mathcal{H}_n$ of arbitrary finite number of Hilbert spaces is defined in a similar way. In quantum mechanics $\mathcal{H}_1 \otimes \cdots \otimes \mathcal{H}_n$ describes a system of n distinguishable particles. In quantum statistical mechanics one considers systems of indistinguishable identical particles - bosons or fermions. In the n-fold multiple tensor product

$$\mathcal{H}^{\otimes n} := \underbrace{\mathcal{H} \otimes \cdots \otimes \mathcal{H}}_{n-times},$$

describing n identical but distinguishable particles, one singles out two subspaces: the symmetric tensor product $\mathcal{H}_+^{(n)}$, describing bosons, and the antisymmetric tensor product $\mathcal{H}_-^{(n)}$, describing fermions (in the case $\mathcal{H} = L^2(\Omega, \mu)$, the former consists of symmetric and the latter – of antisymmetric functions $(\omega_1, \ldots, \omega_n) \to \psi(\omega_1, \ldots, \omega_n)$ in the arguments $\omega_1, \ldots, \omega_n \in \Omega$). Systems of an indefinite (unbounded) number of particles are described by Fock spaces; the *symmetric Fock space*

$$\Gamma_+(\mathcal{H}) = \bigoplus_{n=0}^{\infty} \mathcal{H}_+^{(n)}$$

in the case of bosons and the *antisymmetric Fock space*

$$\Gamma_-(\mathcal{H}) = \bigoplus_{n=0}^{\infty} \mathcal{H}_-^{(n)}$$

for fermions ($\mathcal{H}_+^{(0)} := \mathcal{H}_-^{(0)} := \mathcal{H}^{(0)} \simeq \mathbb{C}$). These spaces carry special representation of the canonical commutation and anticommutation relations respectively, associated with second quantization (see e. g. [35], [66], [41]).

1.3.2 Product States

The tensor product of operators $X_1 \otimes X_2$, where X_j is an operator in $\mathcal{H}_j$ is defined by

$$(X_1 \otimes X_2)(\psi_1 \otimes \psi_2) := X_1\psi_1 \otimes X_2\psi_2.$$

If S_j is a density operator in $\mathcal{H}_j$, then $S_1 \otimes S_2$ is a density operator in $\mathcal{H}_1 \otimes \mathcal{H}_2$, moreover

$$\mathrm{Tr}(S_1 \otimes S_2)(X_1 \otimes X_2) = \mathrm{Tr}\, S_1 X_1 \cdot \mathrm{Tr}\, S_2 X_2 \quad \text{for} \quad X_j \in \mathcal{B}(\mathcal{H}_j).$$

Operators of the form $X \otimes I_2$, where $X \in \mathcal{B}(\mathcal{H}_1)$ and I_2 is the identity operator in $\mathcal{H}_2$, form a subalgebra $\mathcal{B}_1 \subset \mathcal{B}(\mathcal{H}_1 \otimes \mathcal{H}_2)$, isomorphic to $\mathcal{B}(\mathcal{H}_1)$. The formula

$$\mathcal{E}(X_1 \otimes X_2) := X_1 \otimes (\mathrm{Tr}\, S_2 X_2) I_2 \tag{1.34}$$

defines a map $\mathcal{E}$ from $\mathcal{B}(\mathcal{H}_1 \otimes \mathcal{H}_2)$ onto $\mathcal{B}_1$, having the property of a *conditional expectation*

$$\mathbf{E}_S(XY) = \mathbf{E}_S(\mathcal{E}(X)Y) \text{ for } X \in \mathcal{B}(\mathcal{H}_1 \otimes \mathcal{H}_2), Y \in \mathcal{B}_1,$$

for the state $S = S_1 \otimes S_2$. Conditional expectations play a less important part in quantum than in classical probability, since in general the conditional expectation onto a given subalgebra $\mathcal{B}$ with respect to a given state S exists only if $\mathcal{B}$ and S are related in a very special way which in a sense reduces the situation to the classical one; for more detail see Sect. 3.1.3 in Chap. 3.

If S is an arbitrary density operator in $\mathcal{H}_1 \otimes \mathcal{H}_2$, then there is a unique density operator S_1 in $\mathcal{H}_1$ such that

$$\mathrm{Tr}\, S_1 X = \mathrm{Tr}\, S(X \otimes I_2) \text{ for } X \in \mathcal{B}(\mathcal{H}_1). \tag{1.35}$$

(Note that the trace on the left hand side is the trace in $\mathcal{H}_1$ while on the right hand side the trace is taken in $\mathcal{H}_1 \otimes \mathcal{H}_2$.) The same is true for linear combinations of density operators, hence for an arbitrary trace class operator T in $\mathcal{H}_1 \otimes \mathcal{H}_2$. Operator S_1 defined by (1.35) is called *partial trace* of the operator S and denoted by $\mathrm{Tr}_{\mathcal{H}_2} S$. Operation of taking partial trace is similar to calculating marginal distribution of one component of a two-dimensional random variable in classical probability.

Let $S = |\psi\rangle\langle\psi|$ be a pure state in $\mathcal{H}_1 \otimes \mathcal{H}_2$, and S_1, S_2 be its partial states in $\mathcal{H}_1, \mathcal{H}_2$. If S is entangled, that is not product state, the partial states are not pure. Such situation can never happen for classical composite systems. Let us show that S_1, S_2 have the same quantum entropy. Denoting by $\{e_j^1\}$ an orthonormal base of eigenvectors of S_1, and λ_j the corresponding eigenvalues, one has $\psi = \sum_j e_j^1 \otimes h_j$, where $\langle h_j | h_k \rangle = \delta_{jk} \lambda_j$, so that

$$\psi = \sum_j \sqrt{\lambda_j} e_j^1 \otimes e_j^2,$$

where $\{e_j^2\}$ is an orthonormal base of eigenvectors of S_2. Therefore S_1, S_2 have the same nonzero eigenvalues and hence the entropy. This quantity $H(S_1) = H(S_2)$ is zero if and only if S is product state, and serves as a unique measure of entanglement in the pure state. The problem of quantifying entanglement in arbitrary state S is much more complicated; it appears that there are different measures corresponding to different aspects entanglement [32].

1.4 The Problem of Hidden Variables

The problem of hidden variables addresses the possibility of describing quantum mechanics in terms of a classical phase space. Notwithstanding the widely held opinion that such a description is not possible, the construction of hidden variable theories continues (one of the most interesting such attempts is stochastic mechanics [173]). An extensive literature (see for example [86]) is devoted to this problem. Here we content ourselves with a discussion of the essential logical arguments *pro et contra* 'hidden variables'.

1.4.1 Hidden Variables and Quantum Complementarity

From a mathematical point of view, the problem of hidden variables concerns the possibility of establishing correspondence $S \to \hat{S}$ between classical and quantum states, that is between the probability distributions S on a measurable 'phase space' $(\Omega, \mathcal{B}(\Omega))$ and the density operators $\hat{S}$ in the Hilbert space $\mathcal{H}$ of a quantum system, and correspondence $X \rightarrow \hat{X}$ between the random variables X and the observables $\hat{X}$ in $\mathcal{H}$, which reproduce statistical predictions of quantum theory and satisfy some further physically motivated requirements. The minimal requirements, which arise naturally from the concept of statistical model (see Sect. 0.2), are conservation, first, of the functional subordination in the space of observables and, second, of the convex structure in the set of states. A review of the resulting 'no-go' theorems for hidden variables is given in [111]. In particular, the important results of Bell [30] and Kochen and Specker [147] amount to the following:

Proposition 1.4.1. *Let* $\dim \mathcal{H} \geq 3$. *There does not exist an one-to-one map* $\hat{X} \rightarrow X$ *from the set of quantum observables* $\mathfrak{O}(\mathcal{H})$ *into the set of random variables on a measurable space* Ω, *satisfying the following functional condition:*

1. *if* $\hat{X} \rightarrow X$, *then* $f \circ \hat{X} \rightarrow f \circ X$ *for an arbitrary Borel function* f.

Proof. One can assume that $\dim \mathcal{H} < \infty$. Let such a map exist. Then from the property 1. one can derive the following properties

2. *The spectral rule:* $X(\omega) \in \mathrm{Sp}\,\hat{X}$ *for arbitrary* $\omega \in \Omega$.
3. *The finite sum rule:* If the $\hat{X}_j$ are compatible observables and $\hat{X}_j \rightarrow X_j$ then $\sum_j \hat{X}_j \rightarrow \sum_j X_j$.

Let us fix $\omega \in \Omega$ and consider the function of projectors $\mu(\hat{E}) := E(\omega)$, where $\hat{E} \rightarrow E$. From 2. and 3. it follows that μ is a probability measure on $\mathfrak{E}(\mathcal{H})$, assuming only the values 0 and 1. By Gleason's theorem there exists a density operator $\hat{S}$ with $\mu(\hat{E}) = \mathrm{Tr}\,\hat{S}\hat{E}$, but then μ cannot be a two-valued measure.

This improves the famous argument of von Neumann [175], the weakness of which lies in the fact that it required the property 3. for *arbitrary* not just compatible observables. The argument based on additivity of mean values, which was proposed as the motivation for this requirement, essentially excludes hidden variables theories *ab initio* (see e.g. [86], [111]).

The above proof indicates the impossibility of introducing hidden variables by means of a scheme of partial observability, realized, for example, in classical statistical mechanics, where there is a unique correspondence between "macroscopic" observables and certain functions on a phase space. However, it does not exclude the possibility that one and the same quantum observable $\hat{X}$ can be measured by several different methods and hence the correspondence $X \rightarrow \hat{X}$ is not one-to-one. Indeed, in quantum mechanics, there are at least as many different methods of measuring the same observable $\hat{X}$ as there are representations $\hat{X} = f \circ \hat{Y}$ by functions of other observables $\hat{Y}$. If $\hat{X}$ has a multiple eigenvalue, then there exist incompatible observables $\hat{Y}_1$ and $\hat{Y}_2$, such that $\hat{X} = f_1 \circ \hat{Y}_1 = f_2 \circ \hat{Y}_2$. The requirement that the correspondence be one-to-one is then in direct contradiction with quantum complementarity.[2] From this it follows that hidden variables theories must allow the possibility that one and the same quantum observable can have different classical representations. Bell called theories of this kind *contextual*. Similar remarks hold for representations of quantum states by different mixtures of pure states. In fact, the following proposition was proved in [111], [112] by giving an explicit hidden variables construction which preserves the structure of the statistical model.

Proposition 1.4.2. *Let $\mathcal{H}$ be a Hilbert space. There exist a measurable space Ω and maps $X \to \hat{X}$ from a set of random variables onto $\mathfrak{D}(\mathcal{H})$ and $S \to \hat{S}$ from a set of probability measures onto $\mathfrak{S}(\mathcal{H})$ such that*

1. *if $S_j \to \hat{S}_j$ and $\{p_j\}$ is a finite probability distribution, then $\sum_j p_j S_j \to \sum_j p_j \hat{S}_j$;*
2. *if $X \to \hat{X}$ and f is a Borel function then $f \circ X \to f \circ \hat{X}$;*
3. *if $X \to \hat{X}$ and $S \to \hat{S}$, then*

$$\int_\Omega X(\omega) S(d\omega) = \operatorname{Tr} \hat{S}\hat{X}.$$

In the case $\dim \mathcal{H} \geq 3$, the maps $X \to \hat{X}$, and $S \to \hat{S}$ are necessarily not one-to-one. This result shows that complementarity contradicts a classical description of quantum statistics only under the additional requirement of this description being one-to-one (that is, non-contextual). There are contextual hidden variables descriptions of a single quantum system.

1.4.2 Hidden Variables and Quantum Nonseparability

Let us consider now a quantum system consisting of two components described by Hilbert spaces $\mathcal{H}_1$ and $\mathcal{H}_2$. The pure states of the system are represented by unit vectors $\psi \in \mathcal{H}_1 \otimes \mathcal{H}_2$, which are linear combinations (superpositions) of the product vectors $\psi_1 \otimes \psi_2$. If $\psi = \psi_1 \otimes \psi_2$ then both components of the system are in uniquely defined pure states; but if ψ is not a product vector, that is *entangled*, then there are specific correlations between the two components which cannot be modelled by any classical mechanism of randomness. This was pointed out by Bell [30], who showed that even in

---

[2] If $\dim \mathcal{H} = 2$, then observables with multiple spectrum are constants, so that there is no contradiction, and indeed a hidden variable theory satisfying the conditions of the proposition can be constructed in a simple way ([30])

a contextual hidden variables theory it is impossible to satisfy a natural requirement which was called "Einstein locality". We discuss here the related condition of *separability* [111].

Consider the observables

$$\hat{X}_j = \hat{X}_j^{(1)} \otimes \hat{I}^{(2)}, j = 1, \ldots, n,$$
$$\hat{Y}_k = \hat{I}^{(1)} \otimes \hat{Y}_k^{(2)}, k = 1, \ldots, m, \tag{1.36}$$

where $\hat{I}^{(l)}$ is the identity operator in $\mathcal{H}_l$, which satisfy

$$[\hat{X}_j, \hat{Y}_k] = 0, \tag{1.37}$$

that is, each $\hat{X}_j$ is compatible with each $\hat{Y}_k$. Hence, for an arbitrary state $\hat{S}$ in $\mathcal{H}_1 \otimes \mathcal{H}_2$ the quantum correlations $\langle \hat{X}_j, \hat{Y}_k \rangle_{\hat{S}}$ are measurable. The $n \times m$ matrix

$$\mathbf{C} := \left(\langle \hat{X}_j, \hat{Y}_k \rangle_{\hat{S}} \right)_{\substack{j=1,\ldots,n \\ k=1,\ldots,m}}$$

describes the statistical results of $n \cdot m$ different experiments which are in general mutually incompatible.

Proposition 1.4.3. *Let $n, m \geq 2$. There does not exist a measurable space Ω together with maps $S \to \hat{S}$, $X \to \hat{X}$ satisfying the spectral rule and following the separability condition:*
for arbitrary $\hat{S}$ and arbitrary $\hat{X}_1, \ldots, \hat{X}_n, \hat{Y}_1, \ldots, \hat{Y}_m$ of form (1.36) there are X_j and Y_k such that $X_j \to \hat{X}_j$, $Y_k \to \hat{Y}_k$ and

$$\langle \hat{X}_j, \hat{Y}_k \rangle_{\hat{S}} = \int_\Omega X_j(\omega) Y_k(\omega) S(d\omega) \text{ for } j = 1, \ldots, n, \ k = 1, \ldots, m,$$

whenever $S \to \hat{S}$.

Proof. It is sufficient to consider the case $n = m = 2$. Consider the observables $\hat{X}_1, \hat{X}_2, \hat{Y}_1, \hat{Y}_2$ of form (1.36) and such that

$$\|\hat{X}_j\| \leq 1, \|\hat{Y}_k\| \leq 1. \tag{1.38}$$

Assume that maps satisfying the conditions of the proposition exist and let X_j, Y_k be the corresponding random variables on the probability space $(\Omega, \mathcal{B}(\Omega), S)$. By the spectral condition one gets $|X_j(\omega)| \leq 1$, $|Y_k(\omega)| \leq 1$ for all $\omega \in \Omega$, whence

$$X_1(\omega)Y_1(\omega) + X_1(\omega)Y_2(\omega) + X_2(\omega)Y_1(\omega) - X_2(\omega)Y_2(\omega) \le 2 \text{ for } \omega \in \Omega.$$

By averaging over $S(d\omega)$ and using the separability condition, we obtain the *Bell-Clauser-Horn-Shimony (BCHS) inequality*

$$\langle \hat{X}_1, \hat{Y}_1 \rangle_{\hat{S}} + \langle \hat{X}_1, \hat{Y}_2 \rangle_{\hat{S}} + \langle \hat{X}_2, \hat{Y}_1 \rangle_{\hat{S}} - \langle \hat{X}_2, \hat{Y}_2 \rangle_{\hat{S}} \le 2 \qquad (1.39)$$

It remains to give an example of observables $\hat{X}_j$, $\hat{Y}_k$ and a state $\hat{S}$ for which the inequality (1.39) is violated. Consider a system of two particles with spin $\frac{1}{2}$ such that $\dim \mathcal{H}_1 = \dim \mathcal{H}_2 = 2$ (see Sect. 1.1.6). Let $\hat{S} = |\psi\rangle\langle\psi|$ be the pure state of the system given by the vector

$$\psi = \frac{1}{\sqrt{2}} [|\uparrow\downarrow\rangle - |\downarrow\uparrow\rangle] = \frac{1}{\sqrt{2}} [\psi_1(e) \otimes \psi_2(-e) - \psi_1(-e) \otimes \psi_2(e)],$$

where $\psi_j(e)$ is a unit vector in the j-th component describing a completely polarized state with spin direction $e = (0,0,1)$ (see (1.14), (1.15)). Put

$$\hat{X}(a) = \hat{X}^{(1)}(a) \otimes \hat{I}^{(2)}, \quad \hat{Y}(b) = \hat{I}^{(1)} \otimes \hat{X}^{(2)}(b),$$

where $\hat{X}^{(j)}(a)$ is the spin observable in $\mathcal{H}_j$ (see Sect. 1.1.6). The correlations between the spin components have the form

$$\langle \hat{X}(a), \hat{Y}(b) \rangle_{\hat{S}} = \langle \psi | \hat{X}^{(1)}(a) \otimes \hat{X}^{(2)}(b) \psi \rangle = -a \cdot b.$$

Let the vectors a_j, b_k be as in Fig. 1.1, then if $\hat{X}_j = \hat{X}(a_j)$, $\hat{Y}_k = \hat{Y}(b_k)$, the left hand side of (1.39) takes the value $2\sqrt{2}$, which contradicts the inequality and proves the proposition.[3]

In [144] the inequality

$$\left(\hat{X}_1 \hat{Y}_1 + \hat{X}_1 \hat{Y}_2 + \hat{X}_2 \hat{Y}_1 - \hat{X}_2 \hat{Y}_2 \right)^2 \le 4\hat{I} - [\hat{X}_1, \hat{X}_2] \cdot [\hat{Y}_1, \hat{Y}_2],$$

is shown to hold for arbitrary operators satisfying (1.37) and (1.38). From this follow both the BCHS inequality (in the case $[\hat{X}_1, \hat{X}_2] = [\hat{Y}_1, \hat{Y}_2] = 0$) and the bound

---

[3] Bell's work stimulated a number of experiments in which the violation of BCHS type inequalities was confirmed (see, for example [86]).

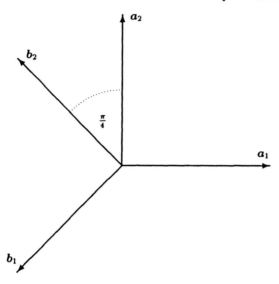

Fig. 1.1. Choice of the vectors a_j and b_k.

$$\|\hat{X}_1\hat{Y}_1 + \hat{X}_1\hat{Y}_2 + \hat{X}_2\hat{Y}_1 - \hat{X}_2\hat{Y}_2\| \leq 2\sqrt{2},$$

from which it is clear that the BCHS inequality is maximally violated in the above example.

Since the components of the composite system can be particles which are spatially separated from one another by a macroscopic distance, a hidden variables theory which describes them must be essentially non-local. [4] In the paper of Summers and Werner [216] it is shown that the situation becomes even sharper with the transition to a local quantum field theory: there the BCHS inequality is maximally violated in an arbitrary entangled state.

1.4.3 The Structure of the Set of Quantum Correlations

In the paper [218] the convex set $\text{Cor}(n, m)$ of *quantum-representable matrices*

$$\mathbf{C} = \left(c_{jk}\right)_{\substack{j=1,\ldots,n \\ k=1,\ldots,m}}$$

was studied, whose elements can be represented as correlations

$$c_{jk} = \left\langle \hat{X}_j, \hat{Y}_k \right\rangle_{\hat{S}}$$

---

[4] For a discussion of nonlocality in stochastic mechanics see the paper [174].

of some quantum observables $\hat{X}_j, \hat{Y}_k$, satisfying (1.37) and (1.38). It turns out that condition (1.36) which is formally stronger (and more natural) than (1.37) defines the same set of correlation matrices C. This is seen from the proof of the following theorem, which gives a transparent geometrical description of the set $\mathrm{Cor}(n, m)$.

Theorem 1.4.1. *The matrix* C *belongs to the set* $\mathrm{Cor}(n, m)$ *if and only if there exist vectors* $a_1, \ldots, a_n, b_1, \ldots, b_m$ *in Euclidean space of dimension* $\min(n, m)$ *such that* $\|a_j\| \leq 1, \|b_k\| \leq 1$ *and* $a_j \cdot b_k = c_{jk}$ *for all* j, k.

We outline the construction which underlies the proof. Let $\mathcal{C}(n)$ be the complex Clifford algebra with n Hermitian generators $X_1, \ldots, X_n$, satisfying

$$X_j^2 = 1, \quad X_j X_k + X_k X_j = 0 \quad \text{for } j, k = 1, \ldots, n, \; j \neq k.$$

Since the elements $X_j \otimes X_j$ of the algebra $\mathcal{C}(n) \otimes \mathcal{C}(n)$ commute and their spectrum consists of ± 1, the number 1 belongs to the spectrum of the element

$$A = \frac{1}{n}(X_1 \otimes X_1 + \cdots + X_n \otimes X_n)$$

One shows it has multiplicity 1. Let π be an irreducible representation of the algebra $\mathcal{C}(n) \otimes \mathcal{C}(n)$. Then there is a vector ψ in the representation space, unique to within a scalar factor, such that $\pi(A)\psi = \psi$. One can show that

$$\langle \psi | \pi(X(a) \otimes X(b)) \psi \rangle = a \cdot b \quad \text{for } a, b \in \mathbb{R}^n,$$

where $X(a) = \sum_{j=1}^n a_j X_j$. The vector ψ defines the state $\hat{S}$ in the faithful representation of the algebra $\mathcal{C}(n) \otimes \mathcal{C}(n)$ such that

$$\langle \hat{X}_j, \hat{Y}_k \rangle_{\hat{S}} = a_j \cdot b_k, \tag{1.40}$$

where $\hat{X}_j = X(a_j) \otimes I, \hat{Y}_k = I \otimes X(b_k)$ satisfy (1.36) and hence (1.37).

From this theorem a description of the extreme points of the set $\mathrm{Cor}(n, m)$ is obtained in [218], in particular, the inequalities defining the set $\mathrm{Cor}(2, 2)$ are found.

Denoting by $\mathrm{Cor}_1(n, m)$ the set of classically-representable matrices C, such that

$$c_{jk} = \int X_j(\omega) Y_k(\omega) S(d\omega),$$

where the X_j and Y_k are random variables satisfying $|X_j(\omega)| \leq 1$ and $|Y_k(\omega)| \leq 1$, we clearly have

$$\mathrm{Cor}_1(n, m) \subsetneq \mathrm{Cor}(n, m).$$

The non-coincidence of these sets is a mathematical expression for quantum nonseparability. In particular, the BCHS inequality provides a hyperplane separating the polyhedron $\mathrm{Cor}_1(2, 2)$ from the quantum-realizable matrix

$$\frac{1}{\sqrt{2}} \begin{pmatrix} 1 & 1 \\ 1 & -1 \end{pmatrix} \in \mathrm{Cor}(2, 2).$$

It is natural to ask by how much $\mathrm{Cor}(n, m)$ exceeds $\mathrm{Cor}_1(n, m)$. Let $k(n, m)$ be the smallest number having the property that

$$\mathrm{Cor}(n, m) \subset k(n, m)\, \mathrm{Cor}_1(n, m).$$

The sequence $k(n, m)$ increases with n and m. It was found in [218] from the geometrical description of the set $\mathrm{Cor}(n, m)$ that

$$k := \lim_{n,m \to \infty} k(n, m) \tag{1.41}$$

coincides with the Grothendieck constant $K_G \leq \frac{\pi}{2\ln(1+\sqrt{2})} \approx 1.782$, known in the theory of normed spaces.

2. Statistics of Quantum Measurements

2.1 Generalized Observables

2.1.1 Resolutions of the Identity

Let $(\mathfrak{X}, \mathcal{B}(\mathfrak{X}))$ be a measurable space. In what follows $\mathfrak{X}$ is often a *standard* measurable space, that is, a Borel subset of a complete separable metric space. Standard measurable spaces of the same cardinality are isomorphic (see, for example, [222] Chap. 5); therefore from a measure theoretical point of view they are equivalent to Borel subsets of the real line $\mathbb{R}$.

A *resolution of the identity* in the Hilbert space $\mathcal{H}$ is a normalized positive operator-valued measure on $\mathcal{B}(\mathfrak{X})$, that is, a set function $M : \mathcal{B}(\mathfrak{X}) \to \mathfrak{B}(\mathcal{H})$ satisfying:

1. $M(B)$ is a positive operator in $\mathcal{H}$ for arbitrary $B \in \mathcal{B}(\mathfrak{X})$;
2. If $\{B_j\}$ is a finite or countable partition of $\mathfrak{X}$ into pairwise disjoint measurable subsets, then
   $\sum_j M(B_j) = I$,
   where the series converges strongly.

If $M(B)^2 = M(B)$ for all $B \in \mathcal{B}(\mathfrak{X})$, then M is an orthogonal resolution of the identity (see Sect. 1.1.3 in Chap. 1). Nonorthogonal resolutions of the identity (on $\mathbb{R}$) appeared first in the work of Carleman (1923) in connection with the problem of spectral decomposition of non-selfadjoint operators, and were studied in detail in the years 1940-1960 (see, for example [7], [199], [34]).

Theorem 2.1.1 (Naimark). *Every resolution of the identity $M : \mathcal{B}(\mathfrak{X}) \to \mathfrak{B}(\mathcal{H})$ in $\mathcal{H}$ can be extended to an orthogonal resolution of the identity, i.e. there exists a Hilbert space $\widetilde{\mathcal{H}} \supset \mathcal{H}$ and an orthogonal resolution of the identity $E : \mathcal{B}(\mathfrak{X}) \to \mathfrak{E}(\widetilde{\mathcal{H}})$ in $\widetilde{\mathcal{H}}$ such that*

$$M(B) = P_{\mathcal{H}} E(B)|_{\mathcal{H}} \quad \text{for all } B \in \mathcal{B}(\mathfrak{X}),$$

where $P_{\mathcal{H}}$ is the projector from $\widetilde{\mathcal{H}}$ onto $\mathcal{H}$. If $\mathcal{H}$ is separable and $\mathfrak{X}$ is standard, then $\widetilde{\mathcal{H}}$ can be chosen to be separable. There exists a minimal extension, unique to within unitary equivalence, characterized by the property that $\{E(B)|_{\mathcal{H}}\psi; B \in \mathcal{B}(\mathfrak{X}), \psi \in \mathcal{H}\}$ is dense in $\widetilde{\mathcal{H}}$.

If there exists a σ-finite measure μ such that $\|M(B)\| \leq C \cdot \mu(B)$ for a fixed $C \in \mathbb{R}$ then

$$M(B) = \int_B P(x)\mu(dx) \text{ for } B \in \mathcal{B}(\mathcal{X}), \qquad (2.1)$$

where $P : \mathcal{X} \to \mathcal{B}(\mathcal{H})$ is a measurable bounded function with values in $\mathfrak{B}(\mathcal{H})$, called the *density* of M with respect to the measure μ (the integral converges in the strong operator topology). If $\dim \mathcal{H} = \infty$, then an orthogonal resolution of the identity cannot have density with respect to a σ-finite nonatomic measure.

Example 2.1.1. In a Hilbert space $\mathcal{H}$ a system $\{e_x ; x \in \mathcal{X}\} \subset \mathcal{H}$ is called *overcomplete* ([146], [35]) if

$$\|\psi\|^2 = \int_{\mathcal{X}} |\langle e_x | \psi \rangle|^2 \mu(dx) \text{ for all } \psi \in \mathcal{H},$$

for some σ-finite measure μ on $\mathcal{B}(\mathcal{X})$, i.e.

$$\int_{\mathcal{X}} |e_x \rangle\langle e_x| \mu(dx) = I.$$

A complete orthogonal system in $\mathcal{H}$ is an example of an overcomplete system, however, in general, the vectors e_x can be non-orthogonal and linearly dependent. Every vector $\psi \in \mathcal{H}$ has a (not necessarily unique) representation

$$\psi = \int_{\mathcal{X}} \langle e_x | \psi \rangle e_x \mu(dx) \qquad (2.2)$$

in terms of the vectors of the overcomplete system. The relation

$$M(B) = \int_B |e_x \rangle\langle e_x| \mu(dx) \qquad (2.3)$$

defines a resolution of the identity with the density $P(x) = |e_x \rangle\langle e_x|$. For it we give an explicit construction of the minimal Naimark dilation (see [56], Chap. 8). Define an orthogonal resolution of the identity E in $\tilde{\mathcal{H}} = L^2(\mathcal{X}, \mu)$ by

$$\left(E(B)f\right)(x) = 1_B \cdot f(x) \text{ for } f \in L^2(\mathfrak{X}, \mu),$$

where 1_B is the indicator of the set $B \in \mathcal{B}(\mathfrak{X})$. From (2.2) and (2.3) it follows that

$$(V\psi)(x) := \langle e_x | \psi \rangle \qquad \text{for } \psi \in \mathcal{H}$$

defines an isometric embedding V of $\mathcal{H}$ into $\widetilde{\mathcal{H}}$, moreover

$$M(B) = V^* E(B)V.$$

The image $V\mathcal{H}$ of $\mathcal{H}$ in $L^2(\mathfrak{X}, \mu)$ is a Hilbert space with reproducing kernel $\mathcal{K}(x, y) := \langle e_x | e_y \rangle$; that is, the projector P from $L^2(\mathfrak{X}, \mu)$ onto $V\mathcal{H}$ is an integral operator with the kernel $\mathcal{K}$.

2.1.2 The Generalized Statistical Model of Quantum Mechanics

This is a separable statistical model (see 0.2) in which the states, like in the standard model, are described by density operators, while observables are resolutions of the identity in the Hilbert space $\mathcal{H}$. The functional calculus for the observables is defined by the relation $(f \circ M)(B) := M(f^{-1}(B))$.

If $\mathfrak{X}$ is a measurable space then a *generalized observable* (respectively, an *observable*) with values in $\mathfrak{X}$ is an arbitrary (respectively orthogonal) resolution of identity M on $\mathcal{B}(\mathfrak{X})$. The probability distribution μ_S^M of a generalized observable M in the state S is defined by

$$\mu_S^M(B) := \operatorname{Tr} SM(B) \text{ for } B \in \mathcal{B}(\mathfrak{X}). \tag{2.4}$$

This definition is justified by the following

Proposition 2.1.1 ([105]). *The map* $S \to \mu_S^M$ *is an affine map from the convex set* $\mathfrak{S}(\mathcal{H})$ *of quantum states into the set of probability measures on* $\mathfrak{P}(\mathfrak{X})$. *Conversely, every affine map from* $\mathfrak{S}(\mathcal{H})$ *into* $\mathfrak{P}(\mathfrak{X})$ *has the form* $S \to \mu_S^M$, *where* M *is a uniquely defined resolution of the identity on* $\mathcal{B}(\mathfrak{X})$.

An affine map transforms mixtures of states into corresponding mixtures of distributions

$$\mu_{\sum_j p_j S_j}^M = \sum_j p_j \mu_{S_j}^M$$

for arbitrary $S_j \in \mathfrak{S}(\mathcal{H})$, $p_j \geq 0$, $\sum_j p_j = 1$, which has a direct interpretation in terms of statistical ensembles. One may say that resolutions of the identity give the most general description of the statistics of the outcomes of quantum measurements which is compatible with the probabilistic interpretation of quantum mechanics.

Using Naimark's theorem, one can prove (see [105], Chap. 1) that for an arbitrary resolution of the identity M in $\mathcal{H}$, there is a Hilbert space $\mathcal{H}_0$, a density operator S_0 in $\mathcal{H}_0$ and an orthogonal resolution of the identity E in $\mathcal{H} \otimes \mathcal{H}_0$, such that

$$\mu_S^M(B) = \mathrm{Tr}(S \otimes S_0)E(B) , \quad \text{for all } B \in \mathcal{B}(\mathcal{X}) \text{ and } S \in \mathfrak{S}(\mathcal{H}). \quad (2.5)$$

Equivalently $M(B) = \mathcal{E}_0(E(B))$, where $\mathcal{E}_0$ is the conditional expectation with respect to the state S_0, defined analogously to (1.34) in Chap. 1. Thus, a resolution of the identity describes the measurement statistics of an ordinary observable in an extension of the initial system, containing an independent auxiliary system in the state S_0. In this way the concept of generalized observable is seen to be compatible with the standard formulation of quantum mechanics.

Example 2.1.2. The minimum uncertainty state vectors (1.30) in Chap. 1 form an overcomplete system in $\mathcal{H} = \mathrm{L}^2(\mathbb{R})$,

$$\iint_{\mathbb{R}^2} |\psi_{x,y}\rangle\langle\psi_{x,y}| \frac{dxdy}{2\pi\hbar} = I,$$

([82], [146]), enabling us to construct a generalized observable with values in $\mathbb{R}^2$ by

$$M(B) = \iint_B |\psi_{x,v}\rangle\langle\psi_{x,v}| \frac{dxdv}{2\pi\hbar} \quad \text{for } B \in \mathcal{B}(\mathbb{R}^2). \quad (2.6)$$

Let us indicate the construction relating M to an approximate joint measurement of the position and momentum of a quantum particle. Let $\mathcal{H}_0 = \mathrm{L}^2(\mathbb{R})$, let P_0 and Q_0 be the canonical observables in $\mathcal{H}_0$ and let $S_0 = |\psi_{0,0}\rangle\langle\psi_{0,0}|$ be the ground state in $\mathcal{H}$. The selfadjoint operators

$$\tilde{Q} = Q \otimes I_0 - I \otimes Q_0, \quad \tilde{P} = P \otimes I_0 + I \otimes P_0 \quad (2.7)$$

in $\mathcal{H} \otimes \mathcal{H}_0$ commute, and hence have joint spectral measure E. Using the method of characteristic functions of Sect. 1.2.4 in Chap. 1 it can be shown that for an arbitrary state S, the probability distribution of the generalized observable (2.6)

$$\mu_S^M(B) = \iint\limits_B \langle \psi_{x,v} | S\psi_{x,v} \rangle \frac{dx\,dv}{2\pi\hbar}$$

satisfies (2.5), that is, coincides with the joint probability distribution of the observables $\tilde{Q}$, $\tilde{P}$ in the state $S \otimes S_0$ (see [105] Chap. 3). Concerning approximate measurements of Q and P see also [56], [94], [221], [42].

2.1.3 The Geometry of the Set of Generalized Observables

The classical analogue of a generalized observable is a transition probability $\Pi(B|\omega)$ from the space of elementary events Ω to the space of outcomes $\mathcal{X}$. Assume that $\mathcal{X}$ is a standard space. Then the relation

$$\Pi(B|\omega) = 1_B(f(\omega)) \quad \text{for } \omega \in \Omega$$

establishes one-to-one correspondence between random variables f with values in $\mathcal{X}$ and *deterministic transition probabilities*, such that $\Pi(B|\omega) = 0$ or 1, that is $\Pi(B|\omega)^2 = \Pi(B|\omega)$. The transition probabilities from Ω to $\mathcal{X}$ form a convex set, whose extreme points are precisely the deterministic transition probabilities (see, for example, [100] Chap. 2).

The relation between observables and generalized observables in the quantum case is significantly more complicated and interesting. Denote by $\mathfrak{M}(\mathcal{X})$ the convex set of all generalized observables with values in $\mathcal{X}$, by $\mathrm{Extr}\,\mathfrak{M}(\mathcal{X})$ its extreme points subset, and by $\mathrm{Conv}\,\mathfrak{M}$ the convex hull of a subset $\mathfrak{M} \subset \mathfrak{M}(\mathcal{X})$. We introduce a natural topology in $\mathfrak{M}(\mathcal{X})$; a sequence $(M^{(n)}) \subset \mathfrak{M}(\mathcal{X})$ converges to M if for an arbitrary state S the sequence of probability measures $\mu_S^{(n)}(B) = \mathrm{Tr}\,SM^{(n)}(B)$ converges in variation to $\mu_S(B) = \mathrm{Tr}\,SM(B)$. Let $\overline{\mathfrak{M}}$ denote the closure of a set $\mathfrak{M} \subset \mathfrak{M}(\mathcal{X})$.

Let us denote by $\mathfrak{M}_0(\mathcal{X})$ the subset of ordinary observables characterized by the condition

$$E(B)^2 = E(B) \quad \text{for } B \in \mathcal{B}(\mathcal{X}),$$

and let $\mathfrak{M}_1(\mathcal{X})$ be the subset of generalized observables M, such that

$$[M(B_1), M(B_2)] = 0 \quad \text{for } B_1, B_2 \in \mathcal{B}(\mathcal{X}).$$

In [95] it is shown that $M \in \mathfrak{M}_1(\mathcal{X})$ if and only if

$$M(B) = \int\limits_{\mathcal{X}_1} \Pi(B|x_1)E(dx_1), \tag{2.8}$$

where E is an observable with values in a space $\mathcal{X}_1$ and $\Pi(B|x_1)$ is a transition probability from $\mathcal{X}_1$ into $\mathcal{X}$. By analogy with classical statistics, observables described by orthogonal resolutions of the identity E can be regarded as *deterministic* (for more detailed discussion see [112]). The observables in $\mathfrak{M}_1(\mathcal{X})$, which are given by commuting resolutions of the identity are *classically randomized* in the sense that each $M \in \mathfrak{M}_1(\mathcal{X})$ is obtained from an ordinary observable by means of transformation (2.8), involving a classical source of randomness. One can also consider a similar transformation for arbitrary generalized observable. The elements of $\mathrm{Extr}(\mathfrak{M}(\mathcal{X}))$ are then generalized observables in which there is no classical randomness caused by a measurement procedure [108]. An interesting example of such an extreme point is given by the non-orthogonal resolution of the identity (2.6). On the other hand, in view of (2.5) every generalized observable $M \in \mathfrak{M}(\mathcal{X})$ may be considered as a *quantum randomized observable* in an extension of the system involving an independent quantum system. The following theorem elucidates the relation between quantum and classical randomizations, implying, in particular, that quantum randomization is more powerful.

Theorem 2.1.2. *Denote by m the number of outcomes in $\mathcal{X}$. If $m = 2$ then $\mathfrak{M}_0(\mathcal{X}) = \mathrm{Extr}(M(\mathcal{X}))$ and $\mathfrak{M}_1(\mathcal{X}) = \overline{\mathrm{Conv}(\mathfrak{M}_0(\mathcal{X}))}$. If $m > 2$ then $\mathfrak{M}_0(\mathcal{X}) \subsetneq \mathrm{Extr}(\mathfrak{M}(\mathcal{X}))$ and $\mathfrak{M}_1(\mathcal{X}) \subsetneq \overline{\mathrm{Conv}\,\mathfrak{M}_0(\mathcal{X})} \subset \mathfrak{M}(\mathcal{X})$, moreover if $\dim \mathcal{H} < \infty$, then the latter inclusion is strict. However, if $\dim \mathcal{H} = \infty$, then $\overline{\mathfrak{M}_0(\mathcal{X})} = \mathfrak{M}(\mathcal{X})$.*

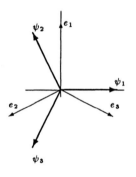

Fig. 2.1. *Information optimum for three vectors on plane.*

Thus the situation is similar to the classical one only in the case of two-valued generalized observables[1]. In this case $\mathfrak{M} = \{M, M'\}$, where $M' = 1 - M$ and as a convex set $\mathfrak{M}(\mathcal{X})$ is isomorphic to the order interval $\{M : M \in \mathfrak{B}_h(\mathcal{H}), 0 \leq M_0 \leq I\}$, whose extreme points coincide with the projectors in $\mathcal{H}$ (see for example, [56] Chap. 2).

To prove that $\mathfrak{M}_0(\mathcal{X}) \neq \mathrm{Extr}\,\mathfrak{M}(\mathcal{X})$ when $m > 2$, it is sufficient to consider the case $m = 3$ and $\dim \mathcal{H} = 2$ (see [106]). Consider the nonorthogonal resolution of the identity

$$M_k = \frac{2}{3}|\psi_k\rangle\langle\psi_k|, \quad k = 1, 2, 3, \qquad (2.9)$$

where ψ_k, $k = 1, 2, 3$, are state vectors of a spin-$\frac{1}{2}$ system (see Sect. 1.1.6 in Chap. 1) that are coplanar and form an equiangular triangle (see Fig. 2.1). The fact that (2.9) is an extreme point can be established either by direct inspection or by using a criterion from Størmer's paper in [71]: a finite resolution of the identity $M = \{M_1, \ldots, M_m\}$ is an extreme point if and only if for arbitrary $X_1, \ldots, X_m \in \mathfrak{B}(\mathcal{H})$ the relation $\sum_{=1}^m E_i X_i E_i = 0$ implies $E_i X_i E_i = 0$, where E_i is the support of M_i, that is, the projector onto

---

[1] Such observables, called "effects" play a central role in the axiomatic approach of Ludwig [161] (see also Kraus [151]).

the orthogonal complement of the null-space of M_i. If $\dim \mathcal{H} < \infty$, then by Caratheodory's theorem, the existence of extreme points not belonging to $\mathfrak{M}_0(\mathcal{X})$ implies $\overline{\mathrm{Conv}(\mathfrak{M}_0(\mathcal{X}))} \neq \mathfrak{M}(\mathcal{X})$.

The proof that $\overline{\mathfrak{M}_0(\mathcal{X})} = \mathfrak{M}(\mathcal{X})$ in the case $\dim \mathcal{H} = \infty$ can be based on Naimark's theorem [106]. Let $M \in \mathfrak{M}(\mathcal{X})$ and let $(P^{(n)})$ be a sequence of finite-dimensional projectors in $\mathcal{H}$, converging strongly to I. Then $M^{(n)}(B) := P^{(n)} M(B) P^{(n)}$ defines a resolution of the identity in the finite-dimensional space $\mathcal{H}^{(n)} := P^{(n)}\mathcal{H}$, which can be extended to an orthogonal resolution of the identity $E^{(n)}$ in the separable Hilbert space $\widetilde{\mathcal{H}}^{(n)} \supset \mathcal{H}^{(n)}$.

Since $\dim \mathcal{H} = \infty$ we may assume that $\widetilde{\mathcal{H}}^{(n)} = \mathcal{H}$. Denoting $\mu_S^{(n)}(B) = \mathrm{Tr}\, S E^{(n)}(B)$, $\mu_S(B) = \mathrm{Tr}\, S M(B)$, we have

$$\mu_S^{(n)}(B) - \mu_S(B) = 2\Re \,\mathrm{Tr}(I - P^{(n)}) S P^{(n)} [E^{(n)}(B) - M(B)]$$

$$+ \mathrm{Tr}(I - P^{(n)}) S (I - P^{(n)}) [E^{(n)}(B) - M(B)],$$

whence

$$\mathrm{var}(\mu_S^{(n)} - \mu_S) \le 6\|(I - P^{(n)}) S\|_1,$$

which proves the last statement of the theorem. Thus, in the case $\dim \mathcal{H} = \infty$ generalized observables are limit points of the set of ordinary observables.

2.2 Quantum Statistical Decision Theory

2.2.1 Optimal Detection

Let there be given a collection of density operators S_θ ($\theta = 1, \ldots, m$) in the Hilbert space $\mathcal{H}$ of observed quantum system, describing possible states of the system. An observer is allowed to make arbitrary quantum measurement on the system. Basing on the outcomes of the measurement he must make a decision in which of the states the observed system is; there is some criterion to compare decision rules and one looks for the optimal one. Mathematically the statistics of a whole decision procedure, including measurement and possible information postprocessing is given by a resolution of the identity $M = \{M_1, \ldots, M_m\}$, called *decision rule*. The probability of a decision $u = 1, \ldots, m$, if the system is in the state S_θ, is then

$$\mu_\theta^M(u) = \mathrm{Tr}\, S_\theta M_u. \tag{2.10}$$

As in the classical statistics, the criterion is given by a functional of the probabilities (2.10), and the problem is to find the extremum of this functional in one or another class of decision rules. In physical problems the quantum system is the information carrier and the state depends on the "transmitted signal" θ. The "receiver" is realized by a quantum measurement, whose outcomes are described by the resolution of identity M in $\mathcal{H}$ [94].

Under the Bayes approach, *a priori* probabilities π_θ for the hypotheses $\theta = 1, \ldots, m$ and the deviation function $W_\theta(u)$ $(\theta, u = 1, \ldots, m)$ are given. The *Bayes risk* is defined by the usual formula

$$\mathcal{R}\{M\} := \sum_{\theta=1}^{m} \pi_\theta \sum_{u=1}^{m} W_\theta(u) \mu_\theta^M(u). \tag{2.11}$$

The decision rule which minimizes $\mathcal{R}\{M\}$, is called *Bayes decision rule*. The case $W_\theta(u) = 1 - \delta_{\theta u}$ and $\pi_\theta = \frac{1}{m}$ is often considered when

$$\mathcal{R}\{M\} = 1 - \mathcal{P}\{M\}$$

where

$$\mathcal{P}\{M\} := \frac{1}{m} \sum_{\theta=1}^{m} \mu_\theta^M(\theta) \tag{2.12}$$

is the average probability of the correct decision. The problem of finding Bayes decision rule is then the question of maximizing $\mathcal{P}\{M\}$, which is a discrete analogue of maximal likelihood in mathematical statistics.

Another important measure of the quality of the decision rule is the *Shannon information*

$$\mathcal{I}\{M\} = \sum_{\theta=1}^{m} \pi_\theta \sum_{u=1}^{m} \mu_\theta^M(u) \log\left(\frac{\mu_\theta^M(u)}{\sum_{\lambda=1}^{m} \pi_\lambda \mu_\lambda^M(u)}\right). \tag{2.13}$$

Denote by $\mathfrak{M}$ the set of all decision rules $\mathfrak{M} = \{M_1, \ldots, M_n\}$, by $\mathfrak{M}_1$ the set of classically randomized rules (that is, such that the M_i commute, i.e. $M_j M_k = M_k M_j$) and by $\mathfrak{M}_0$ the class of deterministic decision rules $(M_j M_k = \delta_{jk} M_j)$. The functionals (2.11) and (2.12) are affine, and (2.13) is convex on the convex set $\mathfrak{M}$; hence they attain their extrema at the subset $\mathrm{Extr}(\mathfrak{M})$. In classical statistics, where $\mathrm{Extr}(\mathfrak{M}) = \mathfrak{M}_0$, the optimal decision rule can be chosen deterministic. As the discussion in Sect. 2.1.3 suggests, in the quantum case the situation is drastically different.

Example 2.2.1 ([95]). Consider the problem of testing of the equiprobable hypotheses

$$S_\theta = |\psi_\theta\rangle\langle\psi_\theta|, \qquad \theta = 1, 2, 3,$$

where ψ_θ are the three equiangular state vectors on the plane (see Fig. 2.1). Then

$$\max_{M \in \mathfrak{M}_0} \mathcal{P}\{M\} = \max_{M \in \mathfrak{M}_1} \mathcal{P}\{M\} = \frac{1}{6}(2 + \sqrt{3}) < \frac{2}{3} = \max_{M \in \mathfrak{M}} \mathcal{P}\{M\},$$

moreover the maximum is attained by the decision rule (2.9). Further,

$$\max_{M \in \mathfrak{M}_0} \mathfrak{J}\{M\} = \max_{M \in \mathfrak{M}_1} \mathfrak{J}\{M\} \approx 0.429 < 0.585 \approx \max_{M \in \mathfrak{M}} \mathfrak{J}\{M\},$$

and the maximum is attained on the decision rule based on the vectors e_1, e_2, e_3, situated as shown on Fig. 2.1 [97].

This and other similar examples (see [58], [202]) show that quantum randomization, unlike classical, can increase the information about the state of observed system. Although this effect occurs only in finite-dimensional Hilbert space (see the last statement of the theorem in the preceding section), it clearly indicates the necessity of using generalized observables.

2.2.2 The Bayes Problem

The Bayes risk (2.11) may be represented in the form

$$\mathcal{R}\{M\} = \operatorname{Tr} \sum_{u=1}^{m} \widehat{W}_u M_u,$$

where $\widehat{W}_u := \sum_{\theta=1}^{m} \pi_\theta W_\theta(u) S_\theta$ is the operator of *a posteriori* deviation. Since $M \to \mathcal{R}\{M\}$ is an affine functional on the convex set $\mathfrak{M}$, the problem of its minimization may be studied using the methods of linear programming.

Theorem 2.2.1 ([100]). *The following duality relation holds:*

$$\min_{M \in \mathfrak{M}} \mathcal{R}\{M\} = \max \left\{ \operatorname{Tr} \Lambda; \Lambda \in \mathfrak{T}(\mathcal{H}), \Lambda \leq \widehat{W}_u; u = 1, \ldots, m \right\}. \quad (2.14)$$

The following conditions are equivalent:

1. $M^0 := \{M_u^0\}$ is Bayes decision rule;
2. there exists $\Lambda^0 \in \mathfrak{T}(\mathcal{H})$ such that

$$\Lambda^0 \leq \widehat{W}_u; \quad (\widehat{W}_u - \Lambda^0) M_u^0 = 0 \text{ for } u = 1, \ldots, m;$$

3. the operator $\Lambda^0 = \sum_{u=1}^{m} \widehat{W}_u M_u^0$ is Hermitian and $\Lambda^0 \leq \widehat{W}_u$ ($u = 1, \ldots, m$).

The operator Λ^0 is the unique solution of the dual problem in the right hand side of (2.14)

The sufficiency of the condition 2. is most frequently used and most easy to prove: for an arbitrary $M = \{M_u\}$

$$\mathcal{R}\{M\} = \operatorname{Tr} \sum_{u=1}^{m} \widehat{W}_u M_u \geq \operatorname{Tr} \Lambda \sum_{u=1}^{m} M_u^0 = \operatorname{Tr} \sum_{u=1}^{m} \widehat{W}_u M_u^0 = \mathcal{R}\{M^0\}.$$

Using this condition it is easy to verify the optimality of the decision rule (2.9) in the example of the preceding section, with $\Lambda^0 = \frac{1}{6}I$.

Example 2.2.2. Let the operators $\widehat{W}_u - \widehat{W}_v$ $(u, v = 1, \ldots, m)$ commute, that is, $\widehat{W}_u = C + \widehat{W}_u^0$ where the $\widehat{W}_u^0$ are commuting operators. Then there exists a selfadjoint operator X and functions $W_u^0(\cdot)$ on $\mathbb{R}$, such that

$$\widehat{W}_u^0 = W_u^0(X).$$

Let $\{\mathcal{X}_k\}$ be a partition of $\mathbb{R}$, such that $W_k^0(x) \leq W_j^0(x)$ for $j \neq k$ and $x \in \mathcal{X}_k$. Put $\Lambda^0 := C + \min_k W_k^0(X)$, $M_k^0 := 1_{\mathcal{X}_k}(x)$. Then condition 2. holds. If $C = 0$, this corresponds to the Bayes decision rule in classical statistics: the rule is deterministic and for each x prescribes the choice of decision u, for which the *a posteriori* deviation W_u is minimal ([94], Chap. IV).

The operator Λ^0 giving the unique solution of the dual problem (2.14) may be regarded as a noncommutative generalization of the function $\min_u \widehat{W}_u$.

Example 2.2.3. The conditions of the preceding example are automatically satisfied in the case of two hypotheses S_0, S_1. For simplicity let us consider the loss function $W_\theta : u \to 1 - \delta_{\theta u}$, so that we are minimizing the mean error. The Bayes decision rule has the form

$$M_0^0 = 1_{(0,\infty)}(\pi_0 S_0 - \pi_1 S_1), \quad M_1^0 = 1_{(-\infty,0]}(\pi_0 S_0 - \pi_1 S_1),$$

and the minimal average error is

$$\mathcal{R}\{M^0\} = \frac{1}{2}\left(1 - \|\pi_0 S_0 - \pi_1 S_1\|_1\right).$$

If S_0, S_1 are pure states with the vectors ψ_0, ψ_1, then this reduces to (see [94])

$$\mathcal{R}\{M^0\} = \frac{1}{2}\left(1 - \sqrt{1 - 4\pi_0\pi_1|\langle\psi_0|\psi_1\rangle|^2}\right).$$

Such bounds given by statistical decision theory provide ideal limits for performance of real measurement procedures and therefore are fundamentally important; however even in the simplest case of two pure states physical realization of the optimal decision rule is by no means an easy problem. An ingenious measurement and information processing device – the Dolinar receiver – was designed for the optimal discrimination between two coherent states of the radiation field, based on photon counting and a specific feedback from the counter to the input field (see [94], [107]). In many other cases the problem of realizing the optimal decision rule still awaits a physical solution. In the general case the optimization equations reduce to a complicated nonlinear problem of geometrical nature.

Let us consider the detection problem for m pure states with linear independent vectors ψ_θ and *a priori* probabilities $\pi_\theta > 0$. We may assume that $\mathcal{H}$ is spanned by the vectors $\psi_\theta, \theta = 1, \ldots, m$. It was shown that in this case the Bayes decision rule has the form

$$M_u = |e_u\rangle\langle e_u|, \quad u = 1, \ldots, m. \tag{2.15}$$

where $\{e_u\}$ is an orthonormal basis in $\mathcal{H}$ (see [94] Chap. 4, [106]). Thus the problem is reduced to finding the orthonormal basis which is the best approximation of the system $\{\psi_u\}$ in the sense of the criterion

$$\mathcal{R}\{M\} = \sum_{\theta=1}^{m} \pi_\theta (1 - |\langle\psi_\theta|e_\theta\rangle|^2). \tag{2.16}$$

It was observed in [103] that in the case of equiprobable pure states the following estimate holds

$$\mathcal{R}\{M\} \leq \frac{1}{m} \sum_{\theta=1}^{m} \|\psi_\theta - e_\theta\|^2.$$

The basis minimizing the right hand side has the form

$$e_k = \sum_{j=1}^{m} a_{kj}\psi_j, \tag{2.17}$$

where $(a_{jk}) := \Gamma^{\frac{1}{2}}$ and $\Gamma := (\langle\psi_j|\psi_k\rangle)$, moreover

$$\min_{e_\theta} \sum_{\theta=1}^{m} \|\psi_\theta - e_\theta\|^2 = \mathrm{Tr}\left(I - \Gamma^{\frac{1}{2}}\right)^2 = 2\,\mathrm{Tr}\left(I - \Gamma^{\frac{1}{2}}\right)$$

(a theorem of M. G. Krein). From this it follows that

$$\min \mathcal{R}\{M\} \leq \frac{1}{m} \operatorname{Tr}\left(I - \Gamma^{\frac{1}{2}}\right)^2. \tag{2.18}$$

The decision rule corresponding to the basis (2.17) is asymptotically optimal in the limit of almost orthogonal states, $\Gamma \to I$, moreover the right hand side in (2.18) gives the first term of the asymptotic. In the equiangular case where $\langle \psi_j | \psi_k \rangle = \gamma$ for $j \neq k$, and $\pi_\theta = \frac{1}{m}$, the basis (2.17) is optimal and one obtains the Yuen-Lax formula ([94], Chap. 6)

$$\min \mathcal{R}\{M\} = \frac{m-1}{m^2} \left(\sqrt{1 + (m-1)\gamma} - \sqrt{1-\gamma}\right)^2.$$

2.2.3 The Quantum Coding Theorem

The issue of the information capacity of quantum communication channel arose in the sixties and goes back to even earlier classical works of Gabor and Brillouin, asking for fundamental physical limits on the rate and quality of information processing. These works laid a physical foundation and raised the question of consistent quantum statistical treatment of the problem. Important steps in this direction were made in the seventies when quantum detection and estimation theory was created, making a mathematical frame for this circle of problems [94], [105]. Dramatic progress has been achieved during recent years, stimulated by the new ideas in quantum information related to the recent development of quantum computing [32], [176].

The simplest model of a quantum communication channel (see [96], [101]) is given by the map $\theta \to S_\theta$, where θ is a "signal" running through the input alphabet $1, \dots, m$, and S_θ are the density operators describing corresponding states of the information carrier (such as radiation field, see [94]). A transmitter produces a probability distribution $\pi = \{\pi_\theta\}$ on the input alphabet (coding), and a receiver performs a decoding described by a resolution of the identity $M = \{M_u\}$, where u runs through the output alphabet $1, \dots, p$ (decoding). The probability of obtaining the output u given the input θ is equal to (2.10). Thus a quantum communication channel may be regarded as an ordinary channel with specific restrictions on the transition probabilities, given implicitly by (2.10). What is the capacity of such a communication channel for transmitting sequences of signals $\theta_1, \dots, \theta_n$?

Consider the information $\mathcal{I}_1\{\pi, M\}$, given by a formula of the type (2.13) (where u runs through 1 to p) and the quantity $C_1 := \max_{\pi, M} \mathcal{I}_1\{\pi, M\}$, where the maximum is over all codings and decodings. In earlier works on quantum communications [83], [154], the quantity

$$\bar{C} = \max_{\pi} \left(H \left(\sum_{\theta=1}^{m} \pi_\theta S_\theta \right) - \sum_{\theta=1}^{m} \pi_\theta H(S_\theta) \right) \qquad (2.19)$$

was used to evaluate the capacity on heuristic grounds, where $H(S)$ is the entropy of the quantum state S. In [98] the *quantum entropy bound*:

$$\mathfrak{I}_1\{\pi, M\} \le H \left(\sum_{\theta=1}^{m} \pi_\theta S_\theta \right) - \sum_{\theta=1}^{m} \pi_\theta H(S_\theta), \qquad (2.20)$$

for an arbitrary coding π and decoding M, was established (see [132] for a chronological survey of relevant results).

Under a similar condition $C_1 < \bar{C}$. However, as the following argument shows, the quantity C_1 should not be considered the true capacity. The genuine definition of capacity must be associated with the limit rate of asymptotically error-free transmission of information. Consider the n-th power of the channel in the space $\mathcal{H}_n := \mathcal{H}^{\otimes n}$, defined by the states $S_v = S_{\theta_1} \otimes \cdots \otimes S_{\theta_n}$, where $v = (\theta_1, \ldots, \theta_n)$ are the possible words of the input alphabet of length n. Let $\mathfrak{I}_n\{\pi, M\}$ and $C_n := \max \mathfrak{I}_n\{\pi, M\}$ be the analogues of $\mathfrak{I}_1\{\pi, M\}$ and C_1 for the n-th power of the channel. The information $\mathfrak{I}_n\{\pi, M\}$ has the property of additivity

$$\max \mathfrak{I}_{n+m}\{\pi^{(n)} \times \pi^{(m)}, M^{(n)} \otimes M^{(m)}\}$$
$$= \max \mathfrak{I}_n\{\pi^{(n)}, M^{(n)}\} + \max \mathfrak{I}_m\{\pi^{(m)}, M^{(m)}\},$$

from which it follows that the sequence (C_n) is superadditive, $C_n + C_m \le C_{n+m}$, and consequently the limit

$$C := \lim_{n\to\infty} \frac{C_n}{n} = \sup_n \frac{C_n}{n}$$

exists. Using the classical Shannon coding theorem, it can be shown that, when $R < C$, there exist codings and decodings of the size $N = [2^{nR}]$, such that the mean error

$$\bar{\lambda}(n, N) = \frac{1}{N} \sum_{j=1}^{N} (1 - \mathrm{Tr}\, S_{v_j} M_j)$$

tends to zero as $n \to \infty$, while when $R > C$, it does not tend to zero for any choice of coding or decoding. This gives the justification for calling the quantity C the (classical) *capacity* of the given quantum communication channel [103].

It must be noted that for the classical memoryless channel the sequence (C_n) is additive and hence $C = \frac{C_n}{n} = C_1$ for all n. In the quantum case it is possible that the sequence is strictly superadditive, $C_n + C_m < C_{n+m}$, moreover

$$C_1 < C,$$

expessing the existence of a paradoxical, from the classical point of view, "memory" in the product of independent quantum channels. An entangled measurement in such a product can give more information than the sum of informations from each component. This fact of course has its roots in the statistical properties of composite quantum systems and is another manifestation of quantum nonseparability.

From (2.20) and the additivity of (2.19) it follows that $C \leq \overline{C}$. It was conjectured in [103] that equality might hold here. In fact, recently it was shown, first for the pure-state channel [91], and then for the general case [132], that

$$C = \overline{C},$$

settling the question of the classical capacity. The inverse inequality $C \geq \bar{C}$ is proved by obtaining an upper bound for the error probability $\bar{\lambda}(n, [2^{nR}])$, which tends to zero as $n \to \infty$ provided $R < \bar{C}$. To get such a bound one has to deal with the minimizations with respect to all possible codings and decodings in (2.2.3). The first minimum is evaluated by using the idea of random coding which goes back to Shannon – an average over an ensemble of random codes is always greater or equal to the minimum. The second minimum is upperbounded by using the special decision rule (2.17). Both these components were already present in the demonstration of strict superadditivity [98], but this was not sufficient to establish the required upper bound. The miraculous improvement of the bound comes by an application of the notion of *typical projection*, a kind of asymptotic equipartition property, which allows us to neglect the "nontypical" eigenvalues of the density operator in question in the limit $n \to \infty$. This notion, which was introduced by Schumacher and Jozsa [143] to implement the quantum analog of data compression, played the key role in obtaining the required $\varepsilon - \delta$ upper bound in the case of pure signal states S_θ [91]. The result was then generalized independently by the author and by Schumacher and Westmoreland to arbitrary signal states (see [132] for a survey).

Later an alternative proof for the case of pure signal states was given, which does not use typical projections, but instead relies upon a large deviation inequality. This allows us to obtain a more precise bound of the form $\delta_c(nR, n) \leq \text{const} \cdot 2^{-nE(R)}$, where $E(R)$ is an estimate for the *reliability function* of the channel [132]. There is still no success in generalizing this proof to the case of arbitrary signal states, although there is a natural conjecture for the possible lower bound for $E(R)$, namely

$$E(R) \geq \max_{0 \leq s \leq 1} \left\{ \max_{\pi} \mu(\pi, s) - sR \right\},$$

where $\mu(\pi, s) = -\log \operatorname{Tr} \left(\sum_{\theta} \pi_{\theta} S_{\theta}^{\frac{1}{1+s}} \right)^{1+s}$.

Example 2.2.4. In the case of the binary quantum channel with pure states ψ_0, ψ_1

$$C = -\left[\left(\frac{1-\varepsilon}{2} \right) \log \left(\frac{1-\varepsilon}{2} \right) + \left(\frac{1+\varepsilon}{2} \right) \log \left(\frac{1+\varepsilon}{2} \right) \right],$$

where $\varepsilon = |\langle \psi_0 | \psi_1 \rangle|$. The maximal amount of information C_1 obtainable with unentangled (product) measurements is attained at the uniform input probability distribution ($\pi = 1/2$) and the corresponding Bayes (maximum likelihood) decision rule given by the orthonormal basis oriented symmetrically with respect to vectors $|\psi_0\rangle$, $|\psi_1\rangle$ (which in this particular case coincides with (2.17)). It is equal to the capacity of a classical binary symmetric channel with the error probability $(1 - \sqrt{1 - \varepsilon^2})/2$, that is

$$C_1 = \frac{1}{2} \left[(1 + \sqrt{1 - \varepsilon^2}) \log(1 + \sqrt{1 - \varepsilon^2}) + (1 - \sqrt{1 - \varepsilon^2}) \log(1 - \sqrt{1 - \varepsilon^2}) \right].$$

In this case $C_1 < C$ and hence $C_n + C_m < C_{n+m}$ for all $0 < \varepsilon < 1$.

In the above setting the classical information was "written" onto quantum states to be transmitted through the physical channel. However the quantum state itself can be regarded as an information resource having properties intrinsically different from any classical source due to entanglement. One of the most fundamental properties of the quantum information is the impossibility of its exact copying (cloning) (see e. g. [228], [176]). The corresponding "quantum information theory" [32] underlies quantum state processing in hypothetical constructs such as quantum computers and in already existing experimental devices such as quantum cryptographic channels [145], [212]. In Section 3.3 of Chap. 3 we shall return to the discussion of channel capacities basing on a general definition of quantum channel.

2.2.4 General Formulation

The detection or hypotheses testing problem is a very special case of the following general scheme. As in the classical statistical decision theory (see e.g. [51]) let there be given a set Θ of values of an unknown parameter θ, a set X of decisions x (usually $X = \Theta$) and a deviation functions $W_\theta(x)$, defining the quality of the estimate x for a given value of parameter θ. The set X is a measurable space (usually standard) and W_θ is bounded from below and measurable in x for fixed $\theta \in \Theta$.

For each value θ there corresponds a density operator S_θ in the Hilbert space $\mathcal{H}$ of the quantum system under investigation, and the *decision rule* is defined by a resolution of the identity M on $\mathcal{B}(\mathcal{X})$. Orthogonal resolutions of the identity describe *deterministic* decision rules. For a given value of the parameter θ and a given decision rule M a decision is made according to the probability distribution

$$\mu_0^M(B) = \operatorname{Tr} S_\theta M(B), \quad B \in \mathcal{B}(\mathcal{X}).$$

The mean deviation, defined by

$$\mathcal{R}_\theta\{M\} = \int_{\mathcal{X}} W_\theta(x)\mu_\theta^M(dx),$$

is an affine functional on the convex set of all decision rules $\mathfrak{M}(\mathcal{X})$, for each $\theta \in \Theta$.

The decision rule called *Bayes* if it minimizes the Bayes risk

$$\mathcal{R}_\pi\{M\} = \int_\Theta \mathcal{R}_\theta\{M\}\pi(d\theta)$$

for a given *a priori* distribution π on Θ, and *minimax* if it minimizes the maximal mean deviation $\max_\theta \mathcal{R}_\theta\{M\}$. Both classical and quantum statistical decision theories are contained in a general scheme in which the states are described by points of an arbitrary convex set $\mathfrak{S}$. Considerable part of Wald's theory can be transferred to this scheme, attaining natural limits of generality [100]. Under minimal requirements for the deviation function the general conditions have been established for the existence of the Bayes [179] and the minimax decision rules, the completeness of the class of Bayes decision rules [37], and an analogue of the Hunt-Stein theorem [105], [37]. Generalizations of the concept of sufficiency were studied in [220], [100] and [188]. Because of the restrictiveness of the concept of conditional expectation, in the quantum case the notion of sufficiency plays a much smaller part than in the classical theory; on the other hand, the properties of invariance under suitable groups of symmetries become more important because of the unique role of symmetries in quantum theory (see Section 2.3).

Using the appropriate concept of integration, necessary and sufficient conditions for optimality together with the duality relation in the Bayes problem with arbitrary $\Theta, \mathcal{X}$, generalizing the theorem of Sect. 2.2.2 were obtained in [100]. Under mild regularity assumptions the Bayes risk is represented as

$$\mathcal{R}_\pi\{M\} = \operatorname{Tr} \int \hat{W}(x)M(dx).$$

The decision rule M^0 is Bayes if and only if there exists a trace-class operator Λ^0 satisfying

$$\Lambda^0 \leq \hat{W}(x), \quad x \in \mathcal{X};$$

$$\int_B (\hat{W}(x) - \Lambda^0) M^0(dx) = 0; \quad B \in \mathcal{B}(\mathcal{X}).$$

The operator Λ^0 is the unique solution of the dual problem of maximizing $\operatorname{Tr} \Lambda$ under the constraints $\Lambda^0 \leq \hat{W}(x), \quad x \in \mathcal{X}$. These conditions in particular enable us to give a complete solution of the multidimensional Bayes problem of estimating the mean value of Gaussian states ([27], [100]), which we illustrate here by an example.

In an estimation problem, $\Theta = \mathcal{X}$ is a finite-dimensional manifold, specifically, a domain in $\mathbb{R}^n$. In this case a deterministic decision rule may be specified by a set of compatible real observables $X_1, \ldots, X_n$ in $\mathcal{H}$. According to 2.1.2, an arbitrary decision rule is specified by a collection of compatible observables $X_1, \ldots, X_n$ in an extension of $\mathcal{H}$ of the form $\mathcal{H} \otimes \mathcal{H}_0$, where $\mathcal{H}_0$ is an auxiliary Hilbert space with a density operator S_0. With such a description, the decision rule is called an *estimate* of the multidimensional parameter $\theta = (\theta_1, \ldots, \theta_n)$. In the quantum estimation problem, as distinct from the classical, the Bayes estimate may turn out to be essentially non-deterministic.

Example 2.2.5. Let $\{S_{\alpha,\beta}; (\alpha, \beta) \in \mathbb{R}^2\}$ be the family of Gaussian states (see Sect. 1.2.4 in Chap. 1) with the characteristic function

$$\operatorname{Tr} S_{\alpha,\beta} \exp(i(Px + Qy)) = \exp\left((\alpha x + \beta y) - \frac{\sigma^2}{2}(x^2 + y^2)\right) \qquad (2.21)$$

for $(x, y) \in \mathbb{R}^2$, where P, Q are canonical conjugate observables, and $\sigma^2 \geq \frac{1}{2}$. Consider the Bayes problem of estimating the two-dimensional parameter $\theta = (\alpha, \beta)$ with the deviation function

$$W_{\alpha,\beta}(\alpha', \beta') := g_1(\alpha - \alpha')^2 + g_2(\beta - \beta')^2,$$

and the Gaussian *a priori* probability density $\frac{1}{2\pi} \exp\left(-\frac{\sigma_0^2}{2}(\alpha^2 + \beta^2)\right)$. The optimum crucially depends on the ratio $\frac{g_1}{g_2}$. If $\frac{g_1}{g_2} \leq (2s^2)^{-2}$ or $\frac{g_1}{g_2} \geq (2s^2)^2$, where $s^2 := \sigma^2 + \sigma_0^2$, then the Bayes estimate is deterministic, that is, given by a pair of commuting selfadjoint operators A, B in $\mathcal{H}$. In the first case $A = \left(\frac{\sigma_0}{s}\right) 2P$, $B = 0$, while in the second $A = 0$, $B = \left(\frac{\sigma_0}{s}\right)^2 Q$. But if $(2s^2)^{-2} \leq \frac{g_1}{g_2} \leq (2s^2)^2$, then the estimates are given by the commuting operators

$$A = k_1(P \otimes I_0) + k_2(I \otimes P_0), \quad B = k_2(Q \otimes I) - k_1(I \otimes Q_0)$$

in $\mathcal{H} \otimes \mathcal{H}_0$ (cf (2.7)), where k_1, k_2 are coefficients that depend nonlinearly on s^2, g_1 and g_2, and the density operator S_0 in $\mathcal{H}_0$ is Gaussian with the characteristic function

$$\text{Tr}\, S_0 \exp(i(P_0 x + Q_0 y)) = \exp\left(-\frac{1}{4}\left(\sqrt{\frac{g_2}{g_1}\frac{k_1}{k_2}}x^2 + \sqrt{\frac{g_1}{g_2}\frac{k_2}{k_1}}y^2\right)\right).$$

The corresponding resolution of the identity in $\mathcal{H}$ differs from (2.7) by a linear change of variables. In contrast to the corresponding classical problem, the dependence of Bayes risk on the weights g_1, g_2 also has essentially nonlinear character.

This example illustrates the new phenomenon of quantum estimation which is not present in the classical statistics: the complementarity between the components of the estimated parameter. The estimates for the components must be compatible observables, and this introduces extra noise in the estimates, provided the weights of the components are of the same magnitude. However, if one component has very low weight as compared to another, the optimal estimation procedure prescribes neglect of the unimportant component and taking of an exact estimate for the important one.

2.2.5 The Quantum Cramér–Rao Inequalities

Let us consider the problem of estimation of a multidimensional parameter $\theta = (\theta_1, \ldots, \theta_k) \in \mathbb{R}^k$ in the family of states $\{S_\theta\}$. A decision rule M is *unbiased* if for all $\theta \in \Theta$

$$\int \cdots \int x_j \mu_\theta^M(dx_1 \ldots dx_k) = \theta_j \text{ for } j = 1, \ldots, k.$$

Assuming finiteness of the second moments we can define the covariance matrix

$$\mathbf{D}_\theta\{M\} := \left(\int \cdots \int (x_i - \theta_i)(x_j - \theta_j)\mu_\theta^M(dx_1 \ldots dx_k)\right)_{i,j=1,\ldots,k}.$$

In classical statistics the covariance matrix of unbiased estimates is bounded from below by the well known Cramér–Rao inequality. The Fisher information matrix which appears in this bound is uniquely determined by the metric geometry of the simplex of "classical states", that is, probability distributions

on the space Ω of the elementary events [51]. In quantum statistics there
are several different inequalities of Cramér–Rao type, reflecting the greater
complexity of the geometry of the state space.

Since the Cramér–Rao inequality has local nature, it is sufficient to assume
that the family of states is defined in a neighborhood of the fixed point θ.
Let us introduce the real Hilbert space $L^2(S_\theta)$, defined as the completion of
the set $\mathfrak{B}_h(\mathcal{H})$ of bounded real observables with respect to the inner product

$$\langle X, Y \rangle_\theta := \Re \operatorname{Tr} Y S_\theta X = \operatorname{Tr} S_\theta X \circ Y, \qquad (2.22)$$

where $X \circ Y := \frac{1}{2}(XY + YX)$ is the Jordan product of X and Y. We assume
that

1. the family $\{S_\theta\}$ is strongly differentiable at θ as a function with values
   in $\mathfrak{T}(\mathcal{H})$;
2. the linear functionals $X \to \operatorname{Tr} \frac{dS_\theta}{d\theta_j} X$ are continuous with respect to the
   inner product (2.22).

Under these conditions, by the theorem of F. Riesz, there exist *sym-metrized logarithmic derivatives* $L_\theta^j \in L^2(S_\theta)$, defined by the equations

$$\operatorname{Tr} \frac{\partial S_\theta}{\partial \theta_j} X = \langle L_\theta^j, X \rangle \quad \text{for } X \in \mathfrak{B}_h(\mathcal{H}). \qquad (2.23)$$

Formally

$$\frac{\partial S_\theta}{\partial \theta_j} = S_\theta \circ L_\theta^j. \qquad (2.24)$$

Then, for an arbitrary decision rule M which has finite second moments and
is *locally unbiased*:

$$\int \cdots \int x_i \frac{\partial \mu_\theta^M}{\partial \theta} (dx_1 \ldots dx_n) = \delta_{ij} \text{ for } i, j = 1, \ldots, k, \qquad (2.25)$$

where $\frac{\partial \mu_\theta^M}{\partial \theta_j}(B) = \operatorname{Tr} \frac{\partial S_\theta}{\partial \theta_j} M(B)$, we have the inequality

$$\mathbf{D}_\theta \{M\} \geq \mathbf{J}_\theta^{-1}. \qquad (2.26)$$

Here $\mathbf{J}_\theta := \left(\langle L_\theta^i, L_\theta^j \rangle \right)_{i,j=1,\ldots,h}$ (a real symmetric matrix) is an analogue of
the Fisher information matrix for the symmetrized logarithmic derivatives.

On the other hand, one can introduce the complex Hilbert spaces $L_\pm^2(S_\theta)$
which are the completions of $\mathfrak{B}(\mathcal{H})$ with respect to the scalar products

$$\langle X, Y\rangle_\theta^+ = \operatorname{Tr} X^* S_\theta Y, \quad \langle X, Y\rangle_\theta^- = \operatorname{Tr} Y S_\theta X^*$$

and define the *right* and *left logarithmic derivatives* $L_\theta^{\pm j}$ as the solutions of the equations

$$\operatorname{Tr} \frac{\partial S_\theta}{\partial \theta_j} X = \langle L_\theta^{\pm j}, X_\theta^\pm \rangle \text{ for } X \in \mathfrak{B}(\mathcal{H}), \tag{2.27}$$

which exist under the same hypotheses 1. , 2. above. Formally

$$\frac{\partial S_\theta}{\partial \theta_j} = S_\theta L_\theta^{+j} = L_\theta^{-j} S_\theta.$$

Then, under the condition (2.25)

$$\mathbf{D}_\theta\{M\} \geq \left(\mathbf{J}_\theta^\pm\right)^{-1}, \tag{2.28}$$

where $\mathbf{J}_\theta^\pm := \left(\langle L_\theta^{\pm i}, L_\theta^{\pm j}\rangle\right)_{i,j=1,\ldots,k}$ are complex Hermitian matrices, and (2.28) is regarded as an inequality for Hermitian matrices.

The formal definition (2.24) of the symmetrized logarithmic derivative and the inequality (2.24) are due to Helstrom, and the inequality (2.28) to Yuen and Lax (see [94] Chap. 8). Other inequalities were obtained by Stratonovich [213]. Rigorous definitions of the logarithmic derivatives and a derivation of the corresponding inequalities is given in Chap. 6 of the book [105]. The L^2 spaces associated with a quantum state are also useful in other contexts. The elements of these spaces can be interpreted as (equivalence classes of) unbounded operators in $\mathcal{H}$ ([105] Chap. 2).

The inequalities (2.26), (2.28) give two different incompatible bounds for $\mathbf{D}_\theta\{M\}$. In the one-dimensional case ($k = 1$) always $J_\theta \leq J_\theta^\pm$, moreover equality holds if and only if $[S_\theta, \frac{dS_\theta}{d\theta}] = 0$. In this case the inequality based on the symmetrized logarithmic derivative turns out to be the optimal ([94] Chap. 8). On the other hand, for the 2-parameter family (2.21) of Gaussian states the inequality (2.26) gives only $\operatorname{Tr} \mathbf{D}_\theta\{M\} \geq 2\sigma^2$, while (2.28) results in $\operatorname{Tr} \mathbf{D}_\theta\{M\} \geq 2\sigma^2 + 1$. The latter bound is attained for unbiased estimates given by operators of the type (2.7) . The inequality (2.28), based on the right (or left) logarithmic derivative is in general better adapted to the estimation problems in which the parameter in the family of the states admits a natural complexification (in the last example $\theta = \alpha + i\beta$). Here again the complementarity between the components comes into play.

2.2.6 Recent Progress in State Estimation

In high precision and quantum optics experiments researchers are already able to operate with elementary quantum systems such as single ions, atoms and photons. This leads to potentially important applications such as quantum communication and quantum cryptography. Quite important is the issue of extracting the maximum possible information from the state of a given quantum system. Thus, in currently discussed proposals for quantum computing, the information is written into states of elementary quantum cells – qubits, and is read off via quantum measurements. From a statistical viewpoint, measurement gives an estimate for the quantum state, either as a whole, or for some of its components (parameters). This brings new interest to the quantum estimation theory, the origin of which dates back to the end of the 1960s – early 1970s (see [94], [105]).

The recent interest in quantum estimation comes on one hand from the aforementioned physical applications, and on the other hand from a natural desire of specialists in classical statistics to enlarge the scope of their science, and to face the new horizons opened by quantum statistics. Quantum estimation theory has several quite distinctive features which make it attractive for an open minded researcher in mathematical statistics. Below we briefly discuss these features, at the same time giving a pilot survey of the most important recent progress in quantum estimation theory.

1) As we have shown in Sect. 2.2.5, in the quantum case estimation problems with multidimensional parameter are radically different from those with one-dimensional parameter. This is due to the non-commutativity of the algebra of quantum observables (random variables), which reflects existence of non-compatible quantities that in principle cannot be measured exactly in one experiment. This sets new statistical limitations to the components of multidimensional estimates, absent in the classical case, and results in essential non-uniqueness of logarithmic derivatives and of the corresponding quantum Cramér–Rao inequalities.

It is well know that in classical statistics the Fisher information gives rise to a Riemannian metric on the set of probability distributions, which is an essentially unique monotone invariant in the cathegory of statistical (Markov) morphisms [51]. It has a natural quantum analog, however the uniqueness no longer holds. The expression

$$d(S_1, S_2) := \sqrt{2(1 - \|\sqrt{S_1}\sqrt{S_2}\|_1)}$$

defines a metric in the set $\mathfrak{S}(\mathcal{H})$ of density operators. In a wider context of von Neumann algebras, this was studied in detail by Araki, Uhlmann and others under the name of *Bures metric* (see a review by Raggio in [193]). If $\{S_\theta\}$ is a family satisfying the conditions 1. , 2. then, as $\Delta\theta \to \theta$,

$$d(S_\theta, S_{\theta+\Delta\theta})^2 \approx \frac{1}{4} \sum_{i,j=1}^{k} \langle L_\theta^i, L_\theta^j \rangle_\theta \Delta\theta_i \Delta\theta_j. \tag{2.29}$$

Thus the Bures metric is locally equivalent to a Riemannian metric defined by the quantum analogue of the Fisher information matrix. Morozova and Chentsov [53] described metrics in $\mathfrak{S}(\mathcal{H})$, where $\dim \mathcal{H} < \infty$, which are monotone invariants in the category of Markov morphisms (that is, dynamical maps, see Sect. 3.1.2 in Chap. 3). In this class the Riemannian metric defined by (2.29) is minimal, while the metric associated with right or left logarithmic derivatives is the maximal one. An example of a metric different both is the Bogoljubov-Kubo-Mori metric playing an important role in quantum statistical mechanics. This question was further investigated in [189] and finally solved in [155], where an exhaustive description of the monotone invariant Riemannian metrics on the set of density operators was given. The implications of this result for quantum estimation are still to be understood.

From a general point of view, studies in the classical geometrostatistics (the term introduced by Kolmogorov) associated with the names of Chentsov, Amari, Barndorff-Nielsen and others lead quite naturally to investigation of the differential-geometric structure of more complicated convex state spaces, such as the set of quantum states (density operators) [219]. In particular, it turns out that the degree of non-commutativity of statistical model and its complex structure are closely related to the Pancharatnam curvature and to Berry's phase, which is well known in quantum mechanics [166].

2) Another fundamental distinction of quantum statistics appears when considering series of independent identical quantum systems, and the asymptotic behavior of the corresponding estimates. In the paper [108] devoted to asymptotics of quantum estimation of a shift parameter, it was found that statistical information in quantum models with independent observations can be strictly superadditive. This property, which is similar to the strict superadditivity of the Shannon information (see Sect. 2.2.3), means that the value of a measure of statistical information for a quantum system consisting of independent components can be strictly greater than the sum of its values for the individual systems. The property of strict superadditivity has deep roots in quantum physics: it is due to existence of entangled (non-separable) measurements over the combined system and is dual to the Einstein-Podolski-Rosen correlations. The phenomenon of strict superadditivity was confirmed also for other important quantum statistical models, such as the full model (see 2) below) [168], [79].

At this point it is appropriate to describe the main concrete models of interest in quantum estimation theory.

1) Parametric models with a group of symmetries. In particular, the models with the shift or rotation parameter are strictly relevant to the issue of canonical conjugacy and nonstandard uncertainty relations, such as time-energy, phase-number of quanta, etc. These models are considered below in Sect. 2.3.5.

2) The full model, in which the multidimensional parameter is the quantum state itself. Although in finite dimensions it is a parametric model with a specific group of symmetries, it deserves to be singled out both because of its importance for physics and of its mathematical features. Especially interesting and mostly studied is the case of qubit state, with the 3-dimensional parameter varying inside the Bloch sphere. Asymptotic estimation theory for the full model in the pure state case was developed in [165], [92]. In this model one sees clearly yet another distinctive feature of quantum estimation: the complexity of the problem increases sharply with transition from pure state to mixed state estimation. In fact, estimation theory for mixed states is an important field to a great extent still open for investigation.

On the other hand, the full model, especially in infinite dimensions, belongs rather to nonparametric quantum mathematical statistics, which is at present also in a stage of development. In this connection we would like to mention the method of homodyne tomography of a density operator in quantum optics [55].

3) Estimation of the mean value of quantum Gaussian states. This is a quantum analog of the classical "signal+noise" problem, however with the noise having quantum-mechanical origin. This model was treated in detail in [99], [100], [105].

2.3 Covariant Observables

2.3.1 Formulation of the Problem

Let G be a locally compact group, acting continuously on the transitive G-space X and let $V : g \to V_g$ $(g \in G)$ be a continuous (projective) unitary representation of the group in the Hilbert space $\mathcal{H}$. A resolution of the identity $M : B \to M(B)$ $(B \in \mathcal{B}(X))$ in $\mathcal{H}$ is called *covariant* under V if

$$V_g^* M(B) V_g = M(g^{-1}B) \quad \text{for } g \in G, B \in \mathcal{B}(X). \tag{2.30}$$

In quantum mechanics X is the space of values of a physical parameter (generalized position) x, having a group G of symmetries (motions). Fix $x_0 \in X$ and a density operator S_0. The relation

$$S_x = V_g S_0 V_g^*, \quad \text{where } x = g x_0, \tag{2.31}$$

describes a transformation of the quantum state, corresponding to the motion g. Namely, if the state S_0 was prepared by certain macroscopic device attached to the origin x_0, the state S_x is prepared by the same device moved to the position $x = g x_0$ (cf. Sect. 1.2.2, Chap. 1). Consider an observable M satisfying the covariance condition (2.30), and let $\mu_x^M(B) := \text{Tr} \, S_x M(B)$ be its probability distribution in the state S_x. Then condition (2.30) is equivalent to the relation

$$\mu^M_{gx_0}(gB) = \mu^M_{x_0}(B) \quad \text{for } g \in G , B \in \mathcal{B}(X), \tag{2.32}$$

holding for an arbitrary state S_0. This means that the statistics of the observable M transforms according to the motion g in the space of the generalized coordinate X (see the example in Sect. 1.2.3, Chap. 1). In this way the covariance condition gives a rule for establishing a correspondence between classical parameters and quantum observables.

Such a correspondence is, of course, far from being one-to-one. Among the covariant generalized observables, those which describe extremally precise measurements of the corresponding parameter are of the main interest. Consider the problem of estimating the parameter $x \in X$ in the family of states (2.31). Let there be given a deviation function $W_\theta(x)$ on the set $X = \Theta$, such that $W_{g\theta}(gx) = W_\theta(x)$. Under the condition (2.30) the mean deviation

$$\mathcal{R}_\theta\{M\} = \int_X W_\theta(x) \mu^M_\theta(dx) \tag{2.33}$$

does not depend on θ. The minimum of the affine functional (2.33) is attained at an extreme point of the convex set $\mathfrak{M}^{G,V}(X)$ of covariant generalized observables. Denote by $\mathfrak{M}^{G,V}_0(X)$ the subset of covariant observables given by orthogonal resolutions of the identity. From the mathematical point of view the correspondence problem amounts to the study of the structure and size of the sets $\mathfrak{M}^{G,V}_0(X)$ and $\mathfrak{M}^{G,V}(X)$. In the general case

$$\mathfrak{M}^{G,V}_0(X) \subsetneq \text{Extr } \mathfrak{M}^{G,V}(X).$$

A number of paradoxes in the standard formulation of quantum mechanics are rooted in the fact that the set $\mathfrak{M}^{G,V}_0(X)$ turns out to be empty. On the other hand, Extr $\mathfrak{M}^{G,V}(X)$ is significantly larger set, in which a generalized quantum observable corresponding to a given classical parameter is usually found.

2.3.2 The Structure of a Covariant Resolution of the Identity

Under special hypotheses concerning G, X, V one can give a complete description of solutions of the covariance equation (2.30), which also throws light on the general case.

Let G be a unimodular group and $G_0 := G/X$ compact. Then there exists a σ-finite invariant measure μ on G and a finite measure ν on X, such that $\nu(B) = \mu(\lambda^{-1}(B))$, where $\lambda : g \to gx_0$ (see e. g. [222]).

Theorem 2.3.1 ([56], [105]). *Let V be a finite-dimensional representation of the group G. For an arbitrary covariant resolution of the identity M there is a positive operator P_0, such that $[P_0, V_g] = 0$, $g \in G_0$ and*

$$M(B) = \int_B P(x)\nu(dx),$$

where the density $P(x)$ is defined by

$$P(gx_0) = V_g P_0 V_g^*. \tag{2.34}$$

Proof. (Sketch). From the identity

$$\int_G \operatorname{Tr} V_g S V_g^* M(B) \mu(dg) = \nu(B) \quad \text{for } S \in \mathfrak{S}(\mathcal{H}),$$

(see [105] Chap. 4),by taking $S = (\dim \mathcal{H})^{-1} I$, we obtain $\operatorname{Tr} M(B) = (\dim \mathcal{H})^{-1} \nu(B)$. Hence there exists a density $P(x)$ with values in $\mathfrak{T}(\mathcal{H})$. The relation (2.34) follows from the covariance condition.

The restriction on P_0, following from the normalization condition $M(\mathcal{X}) = I$, can sometimes be expressed explicitly. It is very simple to describe the set $\mathfrak{M}^{G,V}(\mathcal{X})$ in the case where V is an irreducible square integrable representation $(\dim \mathcal{H} \le +\infty)$ of a unimodular group $G = \mathcal{X}$. From the orthogonality relations for V (see [163]) it follows that for properly normalized μ

$$\int_G V_g S_0 V_g^* \mu(dg) = I$$

for an arbitrary density operator S_0. Thus the formula

$$M(B) = \int_B V_g S_0 V_g^* \mu(dg)$$

establishes one-to-one affine correspondence between $\mathfrak{M}^{G,V}(G)$ and $\mathfrak{S}(\mathcal{H})$. In particular, the extreme points of the set $\mathfrak{M}^{G,V}(G)$ are described by the formula

$$M(B) = \int_B |\psi(g)\rangle\langle\psi(g)| \mu(dg), \tag{2.35}$$

where $\psi(g) = V_g\psi_0$, and ψ_0 is an arbitrary unit vector in $\mathcal{H}$. The family $\{\psi(g); g \in G\}$ forms an overcomplete system, called *generalized coherent states* in the physical literature (ordinary coherent states correspond to an irreducible representation of the CCR and to a special choice of the vector ψ_0 (see Sect. 1.2.4 in Chap. 1).

2.3.3 Generalized Imprimitivity Systems

If M in (2.30) is an orthogonal resolution of the identity then the pair (V, M) is called *imprimitivity system*. This concept was introduced by Mackey (see [163]) and plays an important part in representation theory: a representation V can be extended to an imprimitivity system if and only if it is induced from a subgroup G/X. If M is an arbitrary covariant resolution of the identity then (V, M) is called a generalized imprimitivity system. The following generalization of Naimark's theorem holds.

Theorem 2.3.2 ([46], [205]). *Let G be a second countable locally compact group, and let $(\mathcal{X}, \mathcal{B}(\mathcal{X}))$ be a standard measurable space. Let (V, M) be a generalized system of imprimitivity in $\mathcal{H}$. Then there exist an isometric mapping W from the space $\mathcal{H}$ into a Hilbert space $\widetilde{\mathcal{H}}$ and a system of imprimitivity $(\tilde{V}, E)$ in $\widetilde{\mathcal{H}}$ such that*

$$V_g = W^*\tilde{V}_g W, \quad M(B) = W^* E(B) W.$$

If $\{E(B)W\psi | B \in \mathcal{B}(\mathcal{X}), \psi \in \mathcal{H}\}$ is dense in $\widetilde{\mathcal{H}}$, then $(\tilde{V}, E)$ is unitarily equivalent to the imprimitivity system, extending the representation in $\widetilde{\mathcal{H}} = L^2_{\mathcal{X}}(\mathcal{X}, \mu)$ induced from the subgroup $G/\mathcal{X}$.[2]

As an illustration, consider the pair (V, M), where V is an irreducible square integrable representation, and M is given by (2.35). The desired extension in $\mathcal{H} = L^2(G, \mu)$ is given by

$$\tilde{V}_g f(x) = f(g^{-1}x), \quad E(B)f(x) = 1_B(x)f(x),$$

moreover the embedding W acts by $W\psi(g) := \langle\psi(g)|\psi\rangle$. The subspace $W\mathcal{H} \subset L^2(G, \mu)$ is associated with the reproducing kernel $K(g, g') := \langle\psi(g)|\psi(g')\rangle = \langle\psi_0|V(g^{-1}g')\psi_0\rangle$. The connection between generalized coherent states and induced representations was studied in detail in [205]. In the general case in [47] it was shown that M has a bounded density $P(x)$ with respect to the quasi-invariant measure μ on $\mathcal{X}$ if and only if the subspace $W\mathcal{H} \subset L^2_{\mathcal{X}}(\mathcal{X}, \mu)$ is a reproducing kernel Hilbert space (with values in $\mathcal{B}(\mathcal{X})$).

---

[2] $L^2_{\mathcal{X}}(\mathcal{X}, \mu)$ denotes the Hilbert space of functions on $\mathcal{X}$ with values in the Hilbert space $\mathcal{X}$, square integrable with respect to the measure μ.

2.3.4 The Case of an Abelian Group

The case when $G = X$ is an Abelian locally compact group is interesting particularly in connection with canonical conjugacy in quantum mechanics. A complete description of covariant resolutions of the identity for an arbitrary (continuous) representation V is given in terms of the Fourier transform; here the values of the "density" $P(x)$ in general are unbounded non-closable positive-definite forms. The results discussed below can be obtained either by direct methods of harmonic analysis or by using the extension theorem of the preceding section (see [109], [110]). The generalization to non-Abelian groups of type I is given in [113]. We denote by $\hat{G}$ the dual group, and by $d\hat{g}$ the Haar measure in $\hat{G}$.

Proposition 2.3.1. *The set of covariant generalized observables $\mathfrak{M}^{G,V}(\mathcal{G})$ is nonempty if and only if the spectrum of V is absolutely continuous with respect to $d\hat{g}$.*

If this condition is satisfied then V decomposes into a direct integral of the factor representations, that is

$$\mathcal{H} = \int_{\Lambda} \oplus \mathcal{H}(\lambda) d\lambda,$$

where Λ is a measurable subset of $\hat{G}$ and $\{\mathcal{H}(\lambda); \lambda \in \Lambda\}$ is a measurable family of Hilbert spaces with $\dim \mathcal{H}(\lambda) > 0$ for almost all $\lambda \in \Lambda$, moreover

$$V_g \psi = \int_{\Lambda} \oplus \overline{\lambda(g)} \psi(\lambda) d\lambda, \quad \text{if } \psi = \int_{\Lambda} \oplus \psi(\lambda) d\lambda.$$

Here $\lambda(g)$ is the value of the character $\lambda \in \hat{G}$ on the element $g \in G$.

The next proposition, which should be compared to the proposition 2.3.1, follows from Mackey's imprimitivity theorem, which generalizes the Stone-von Neumann uniqueness theorem (see Sect. 1.2.3 in Chap. 1).

Proposition 2.3.2. *The set of ordinary covariant observables $\mathfrak{M}_0^G(G, V)$ is nonempty if and only if $\Lambda = \hat{G}$ to within a set of zero measure and $\dim \mathcal{H}(\lambda) = $ const for almost all $\lambda \in \hat{G}$.*

We call a family $\{P(\lambda, \lambda'); \lambda, \lambda' \in \Lambda\}$ a *kernel* if the $P(\lambda, \lambda')$ are contraction operators from $\mathcal{H}(\lambda')$ into $\mathcal{H}(\lambda)$, and the complex valued function $\langle \varphi(\lambda)|P(\lambda, \lambda')\psi(\lambda')\rangle_\lambda$ is measurable with respect to $d\lambda \times d\lambda'$ for arbitrary $\varphi = \int \oplus \varphi(\lambda) d\lambda, \psi = \int \oplus \psi(\lambda) d\lambda \in \mathcal{H}$ (here $\langle \cdot|\cdot\rangle_\lambda$ denotes the scalar product in $\mathcal{H}(\lambda)$). The kernel is *positive definite* if

$$P(\varphi, \varphi) := \iint\limits_{\Lambda\ \Lambda} \langle \varphi(\lambda)|P(\lambda, \lambda')\varphi(\lambda')\rangle_\lambda d\lambda d\lambda' \geq 0$$

for all $\varphi \in \mathcal{H}_1 := \{\varphi; \int\limits_\Lambda \||\varphi(\lambda)\||_\lambda d\lambda < \infty\}$. Note that the positive form $P(\varphi, \varphi)$ need not be closable, i.e. associated with any selfadjoint operator. For such kernels the diagonal value $P(\lambda, \lambda)$ is well defined up to equivalence (see [109]). Let us denote $\hat{1}_B(\lambda) := \int\limits_B \lambda(g)dg$, where dg is a Haar measure in G.

Theorem 2.3.3. *The relation*

$$\langle \varphi|M(B)\psi\rangle = \int\limits_B P(V_g^*\varphi, V_g^*\varphi)dg \qquad (2.36)$$

$$= \iint\limits_{\Lambda\ \Lambda} \langle \varphi(\lambda)|P(\lambda, \lambda')\varphi(\lambda')\rangle_\lambda \hat{1}_B(\lambda - \lambda')d\lambda d\lambda',$$

where B runs through the compact subsets of G and $\varphi \in \mathcal{H}_1$, establishes a one-to-one correspondence between the covariant resolutions of the identity in $\mathcal{H}$ and the quadratic forms $P(\varphi, \varphi)$ on $\mathcal{H}_1$ given by positive definite kernels $\{P(\lambda, \lambda')\}$, such that $P(\lambda, \lambda) = I_\lambda$ (the unit operator in $\mathcal{H}(\lambda)$).

This theorem reduces description of the set Extr $M^{G,V}(G)$ to finding the extreme points of the convex set of positive definite kernels $\{P(\lambda, \lambda')\}$ satisfying the condition $P(\lambda, \lambda) = I_\lambda$. In full generality this problem is not solved even for finite Λ. It is however possible to distinguish a subclass of the set Extr $\mathfrak{M}^G(G, V)$, which is important for quantum mechanical applications.

For simplicity we further limit ourselves to the case where $\dim \mathcal{H}(\lambda) = $ const for all $\lambda \in \Lambda$. Denote by $\mathfrak{M}_c^G(G, V)$ the class of covariant resolutions of the identity, the kernels of which satisfy

$$P(\lambda, \lambda')P(\lambda', \lambda'') = P(\lambda, \lambda'') \text{ for arbitrary } \lambda, \lambda', \lambda'' \in \Lambda.$$

In the case $\Lambda = G$ this class coincides with $\mathfrak{M}_0^G(G, V)$, on the other hand $\mathfrak{M}_c^G(G, V) \subset \text{Extr }\mathfrak{M}^G(G, V)$, moreover equality holds only if Λ consists of two points. All the elements $\mathfrak{M}_c^G(G, V)$ are obtained one from another by *gauge transformations*

$$M'(B) = U^*M(B)U, \qquad (2.37)$$

where $U = \int\limits_\Lambda \oplus U(\lambda)d\lambda$ is a decomposable unitary operator in $\mathcal{H} = \int\limits_\Lambda \oplus\mathcal{H}(\lambda)d\lambda$.

If $\mathcal{H}$ is realized as $L_\mathcal{K}^2(\Lambda, d\lambda)$, where $\mathcal{K} := \mathcal{H}(\lambda)$ for $\lambda \in \Lambda$, then in the class $\mathfrak{M}_c^G(G, V)$ there is a distinguished resolution of the identity M_c, defined by the kernel $P(\lambda, \lambda') \equiv I_\mathcal{K}$ (the unit operator in $\mathcal{K}$) for which

$$\langle\psi|M_c(B)\varphi\rangle = \int\limits_{A}\int\limits_{A} \langle\psi(\lambda)|\varphi(\lambda')\rangle_{\mathcal{K}}\,\hat{1}_B(\lambda - \lambda')d\lambda d\lambda'. \qquad (2.38)$$

Each $M \in \mathfrak{M}_c^G(G, V)$ has a kernel of the form $U(\lambda)^*U(\lambda')$, where $\{U(\lambda)\}$ is a measurable family of unitary operators in $\mathcal{K}$.

2.3.5 Canonical Conjugacy in Quantum Mechanics

Let x be a one dimensional parameter so that $G = X$ is the real line $\mathbb{R}$ (the case of a displacement parameter) or the unit circle $\mathbb{T}$ (the case of a rotational parameter), and let $x \to V_x = e^{-ixA}$ be a unitary representation of the group G in $\mathcal{H}$. The spectrum Λ of the operator A is contained in the dual group $\hat{G}$, which is identified with $\mathbb{R}$ in the case $G = \mathbb{R}$ and with the group of integers $\mathbb{Z}$ when $G = \mathbb{T}$.

The covariance relation for a generalized observable M has the form

$$V_x^* M(B)V_x = M(B - x) \text{ for } B \in \mathcal{B}(G),\ x \in G, \qquad (2.39)$$

where $B - x := \{y; y + x \in B\}$, and in the case $G = \mathbb{T}$ addition is modulo 2π. We introduce the operators

$$U_y := \int\limits_{G} e^{iyx} M(dx) \text{ for } y \in \hat{G}. \qquad (2.40)$$

Then (2.39) reduces to the Weyl relation (see Sect. 1.2.3 in Chap. 1)

$$U_y V_x = e^{ixy}V_x U_y \text{ for } x \in G,\ y \in \hat{G}, \qquad (2.41)$$

in which, however, the operator U_y, is in general non-unitary. In this sense the generalized observable M is canonically conjugate to the observable A.

For the generalized canonical pair (A, M) the following uncertainty relation holds [108]

$$\Delta_S^M(y) \cdot \mathbf{D}_S(A) \geq \frac{1}{4} \text{ for } y \in \hat{G}, \qquad (2.42)$$

where $\Delta_S^M(y) := y^{-2}(|\operatorname{Tr} SU_y|^{-2} - 1)$, $y \neq 0$, is a functional measure of uncertainty of the covariant generalized observable M in the state S. If $G = \mathbb{R}$ and M has finite variance $\mathbf{D}_S(D)$, then $\lim_{y\to 0} \Delta_S^M(y) = \mathbf{D}_S(M)$, so that from (2.42) there follows the generalization of the Heisenberg uncertainty relation

$$\mathbf{D}_S(M)\mathbf{D}_S(A) \geq \frac{1}{4}.$$

For a rotational parameter $(G = \mathbb{T})$ variance is not an adequate measure of uncertainty, and the inequality (2.42) must be considered as final. Different uncertainty relations for angular variables have been discussed in the reviews [44] and [64]. Note that a generalized observable representing the angle of rotation always exists, since the conditions of proposition 2.3.1 of the previous section are automatically satisfied (with $\widehat{G} = \mathbb{Z}$). On the other hand, the conditions of proposition 2.3.2 cannot be satisfied if $\dim \mathcal{H} < \infty$ (as for systems of finite spin) and in these cases an ordinary observable representing the angle of rotation does not exist.

Of the main interest are covariant observables having the minimal uncertainty. The following result describes them in the case of a pure initial state.

Theorem 2.3.4 ([108]). *Let $S = |\psi\rangle\langle\psi|$ be a pure state, then*

$$\min_{M \in \mathfrak{M}^G(G,V)} \Delta_S^M(y) = y^{-2}\left(\left(\int_{\dot{G}} \|\psi(y')\|\|\psi(y'+y)\|dy'\right)^{-2} - 1\right). \tag{2.43}$$

The minimum is achieved on the covariant observable M^* of the class $\mathfrak{M}_c^G(G,V)$, which is given by a kernel $P^*(y,y')$ such that

$$\frac{P^*(y,y')\psi(y')}{\|\psi(y')\|} = \frac{\psi(y)}{\|\psi(y)\|} \quad \text{for } y,y' \in \Lambda.$$

In particular, for $G = \mathbb{R}$, (2.43) implies

$$\min_{M \in \mathfrak{M}^{\mathbb{R}}(\mathbb{R},V)} \mathbf{D}_S(M) = \int_{\mathbb{R}} \left(\frac{d}{dy}\|\psi(y)\|\right)^2 dy, \tag{2.44}$$

where $\psi(y)$ are the components of the vector ψ in the representation diagonalizing the operator A. The quantities (2.43) , (2.44) give intrinsic measures of uncertainty of the parameter x in the state S.

In this way, the requirements of covariance and minimal uncertainty (with respect to pure states) define a canonically conjugate generalized observable, which is unique to within a gauge transformation (2.37). Note that the same arbitrariness also remains in the standard formulation of quantum mechanics, since in the case $\Lambda = \widehat{G}$ the class $\mathfrak{M}_c^G(G,V)$ coincides with the class of covariant observables $\mathfrak{M}_0^G(G,V)$. The quantum-mechanical description of the non-standard canonical pairs such as angle-angular momentum, phase-number of quanta and time-energy from this viewpoint is given in [102], [104], [105]. Let us briefly consider here the case of the time observable.

Example 2.3.1. Consider a quantum system with positive Hamiltonian H. The representation $t \to V_t := e^{iHt/\hbar}$ of the group of time translations does not satisfy the conditions of proposition 2.3.2 of the previous section since $\Lambda \subset \mathbb{R}_+$. Hence in the standard model there is no covariant time observable. Assume, for simplicity, that $\Lambda = \mathbb{R}_+$ and that the spectrum of H is homogeneous (that is, it has a constant multiplicity for almost all $\lambda \in \Lambda$). Then H is unitarily equivalent to the operator of multiplication by λ in the Hilbert space $\mathcal{H} = L^2_{\mathcal{K}}(\mathbb{R}_+)$ of square-integrable functions $\psi = [\psi(\lambda)]$ on $\mathbb{R}_+$ with values in Hilbert space $\mathcal{K}$. Covariant generalized observables of the class $\mathfrak{M}^{\mathbb{R}}_c(\mathbb{R}, V)$ are equivalent to within a gauge transformation (2.37) to the observable M_c, defined by (2.38), that is,

$$\langle \psi | M_c(B) \varphi \rangle = \iint_{\mathbb{R}_+ \mathbb{R}_+} \langle \psi(\lambda) | \varphi(\lambda') \rangle_{\mathcal{K}} \int_B e^{-i(\lambda'-\lambda)\tau} \frac{d\tau}{2\pi} d\lambda d\lambda' \qquad (2.45)$$

The operators (2.40) are given by

$$U_y = \begin{cases} P_y; \ y \geq 0 \\ P_y^*; \ y < 0 \end{cases},$$

where $\{P_y\}$ is the contractive semigroup of one-sided shifts in $L^2_{\mathcal{K}}(\mathbb{R}_+)$:

$$P_y \psi(\lambda) = \psi(\lambda + y); \quad y \geq 0.$$

The generator of this semigroup is iT, where T is the maximal symmetric (but not selfadjoint) operator

$$T = \frac{1}{i} \frac{d}{d\lambda},$$

$$\mathcal{D}(T) = \{\psi; \ \psi \text{ is absolutely continuous}, \ \psi(0) = 0, \ \int_{\mathbb{R}_+} \|\frac{d}{d\lambda}\psi(\lambda)\|^2_{\mathcal{K}} d\lambda < \infty\}.$$

The resolution of the identity M_c is a generalized spectral measure of this operator in the sense of [7].

The minimal Naimark dilation of the resolution of the identity M_c in the space $\tilde{\mathcal{H}} = L^2_{\mathcal{K}}(\mathbb{R})$ is obtained replacing $\mathbb{R}$ by $\mathbb{R}_+$ in (2.45) and is the spectral measure of the selfadjoint operator $i^{-1}\frac{d}{d\lambda}$ in $L^2_{\mathcal{K}}(\mathbb{R})$.

For the free particle of mass m one can use the momentum rather than energy representation to compute

$$T = m \sum_{j=1}^{3} \frac{P_j}{|P|^2} \circ Q_j$$

with the maximal domain on which $\|T\psi\| < \infty$ [102]. This agrees with the expression used in scattering theory for evaluation of the collision time.

2.3.6 Localizability

Let W be a unitary representation in the Hilbert space $\mathcal{H}$, of the group describing the kinematics of a given quantum system (that is, the universal covering of the Galilei group in the non-relativistic case or the Poincaré group in the relativistic case), and let U be the restriction of W to the universal covering G of the group of Euclidean transformations $g : x \to Ax + b :$ $\mathbb{R}^3 \to \mathbb{R}^3$. The system is said to be *Wightman localizable* if there exists an orthogonal resolution of the identity E in $\mathcal{H}$, over the σ-algebra $\mathcal{B}(\mathbb{R}^3)$ of Borel subsets of the coordinate space $\mathbb{R}^3$, satisfying the Euclidean covariance relation

$$U_g^* E(B) U_g = E(g^{-1}B) \text{ for } g \in G, \ B \in \mathcal{B}(\mathbb{R}^3).$$

Under this condition, for an arbitrary open domain $B \subset \mathbb{R}^3$ there is a state vector ψ such that $\langle \psi | E(B) \psi \rangle = 1$, that is, the probability that the system is found in the domain B is equal to 1. For a localizable system, there exist compatible position observables

$$Q_j = \iiint x_j E(dx_1 dx_2 dx_3); \quad j = 1, 2, 3, \tag{2.46}$$

covariant under the Euclidean transformations.

Using the fact that (U, E) is a system of imprimitivity, one may show that all massive particles and massless relativistic particles with zero chirality are Wightman localizable. The massless particles with non-zero chirality (photons and neutrinos) turn out to be non-localizable [222]. This conclusion contradicts the experimental localizability of photon; furthermore, as Hegerfeldt showed (see, for example [9]), this notion of localizability is inconsistent with the requirement of causality in relativistic dynamics.

These difficulties are removed if we allow arbitrary resolutions of the identity in the definition of localizability. Non-orthogonal resolutions of the identity M, describing the localization of a photon were given, in particular, in the work of Kraus in [221] and the author [104]. A complete classification of the corresponding systems of imprimitivity, under the additional condition of covariance with respect to similarity transformations, including the characterization of extreme points, was given in [45]. The massless relativistic particles prove to be *approximately localizable* in the sense that $\sup_{\|\psi\|=1} \langle \psi | M(B) \psi \rangle = 1$ for an arbitrary open domain $B \subset \mathbb{R}^3$. The position observables, determined by relations (2.46) are selfadjoint but noncommuting operators and therefore do not have a joint spectral measure.

In a number of works, of which there is a review in the papers [9], the idea is developed of stochastic localizability in the phase space. The results of these works also indicate that the generalized statistical model of quantum mechanics provides at least the possibility of softening the well-known conflict between relativistic covariance and the non-locality of quantum mechanical description.

3. Evolution of an Open System

3.1 Transformations of Quantum States and Observables

The dynamics of isolated quantum system, described by one-parameter group of unitary operators, is reversible in time. The evolution of an open system, subject to exterior influences, whether it be the process of establishing equilibrium with the environment or interaction with the measuring apparatus, reveals features of irreversibility. Such irreversible changes are described mathematically by completely positive maps.

3.1.1 Completely Positive Maps

Let $\mathfrak{A} \subset \mathfrak{B}(\mathcal{H})$ be a C^*-*algebra* of operators, that is, a subspace of $\mathfrak{B}(\mathcal{H})$, closed under algebraic operations, involution and passage to the limit with respect to the operator norm (see [206], [40]). Denote by $\mathfrak{M}_n$ the algebra of complex $n \times n$ -matrices. A linear map $\Phi : \mathfrak{A} \to \mathfrak{B}(\mathcal{K})$, where $\mathcal{K}$ is a Hilbert space, is called *positive* if from $X \in \mathfrak{A}$, $X \geq 0$, it follows that $\Phi(X) \geq 0$ and *completely positive* if for all $n \geq 1$ the linear map Φ_n from the C^*-algebra $\mathfrak{A} \otimes \mathfrak{M}_n$ to the C^*-algebra $\mathfrak{B}(\mathcal{K}) \otimes \mathfrak{M}_n$ defined by $\Phi_n(X \otimes Y) := \Phi_n(X) \otimes Y$, is positive. In other words, for an arbitrary matrix $(X_{jk})_{j,k=1,...,n}$ with elements $X_{jk} \in \mathfrak{A}$, which is positive definite in the sense that $\sum_{j,k=1}^{n} \langle \phi_j | X_{jk} \phi_k \rangle \geq 0$ for arbitrary family $\{\phi_j\} \subset \mathcal{H}$, the matrix $(\Phi(X_{jk}))_{j,k=1,...,n}$ is also positive definite. There is yet another equivalent definition: for arbitrary finite families $\{X_j\} \subset \mathfrak{A}$ and $\{\psi_j\} \subset \mathcal{K}$

$$\sum_{j,k} \langle \psi_j | \Phi(X_j^* X_k) \psi_k \rangle \geq 0. \tag{3.1}$$

For a positive map, the following *Kadison-Schwarz inequality* holds

$$\Phi(X)^* \Phi(X) \leq \|\Phi\| \Phi(X^* X) \tag{3.2}$$

for all $X \in \mathfrak{A}$ such that $X^* X = X X^*$ (such operators are called *normal*). If Φ is completely positive, then this inequality is satisfied for all $X \in \mathfrak{A}$. An example of a positive, but not completely positive map (related to time inversion, see Sect. 1.2.1 in Chap. 1) is the transposition in $\mathfrak{M}_n$. If Φ is positive and $\mathfrak{A}$ or $\Phi(\mathfrak{A})$ is commutative, then Φ is completely positive. Thus the property of complete positivity appears nontrivially only in the noncommutative case.

A map $\pi : \mathfrak{A} \to \mathfrak{B}(\mathcal{K})$ is called a *-homomorphism*, if it preserves the algebraic operations and involution.

Theorem 3.1.1 (Stinespring). *Let $\mathfrak{A} \subset \mathfrak{B}(\mathcal{H})$ be a C^*-algebra with identity and let $\Phi : \mathfrak{A} \to \mathfrak{B}(\mathcal{K})$ be a linear map. Φ is completely positive if and only if it admits the representation*

$$\Phi(X) = V^* \pi(X) V, \tag{3.3}$$

*where V is a bounded linear map from $\mathcal{K}$ into a Hilbert space $\widetilde{\mathcal{K}}$, and π is a *-homomorphism of $\mathfrak{A}$ in $\mathfrak{B}(\widetilde{\mathcal{K}})$.*

There exists a minimal representation (3.3) (unique to within a unitary equivalence) characterized by the property: the subspace $\{\pi(X)V\psi : X \in \mathfrak{A}, \psi \in \mathcal{K}\}$ is dense in $\widetilde{\mathcal{K}}$.

The proof of the direct statement is a generalization of the *Gelfand-Naimark-Segal (GNS) construction* which corresponds to the case of a positive linear functional on $\mathfrak{A}$ (that is $\dim \mathcal{K} = 1$) [206], [40]. On the algebraic tensor product $\mathfrak{A} \otimes \mathcal{K}$ we define an inner product, such that

$$\langle X \otimes \varphi | Y \otimes \psi \rangle = \langle \varphi | \Phi(X^* Y) \psi \rangle_{\mathcal{K}} \quad \text{for } X, Y \in \mathfrak{A}, \varphi, \psi \in \mathcal{K}.$$

Positive definiteness of this inner product follows from (3.1). Let $\widetilde{\mathcal{K}}$ be the Hilbert space obtained by factorization and completion of $\mathfrak{A} \otimes \mathcal{H}$ with respect to this product. The relations

$$\pi[X](Y \otimes \psi) = XY \otimes \psi$$
$$V\varphi = I \otimes \varphi$$

define in $\widetilde{\mathcal{K}}$ the objects V, π, satisfying (3.3).

Remark 3.1.1. If $\mathcal{X}$ is a metrizable compact set and M is a resolution of the identity in $\mathcal{K}$ on the σ-algebra of the Borel subsets $\mathcal{B}(\mathcal{X})$, then the map

$$f \to \int_{\mathcal{X}} f(x) M(dx) \text{ for } f \in C(\mathcal{X})$$

from the C^*-algebra $C(\mathcal{X})$ of continuous complex functions on $\mathcal{X}$ into $\mathfrak{B}(\mathcal{K})$ is (completely) positive. In this case Stinespring's theorem gives the Naimark's extension for M, since *-homomorphism of $C(\mathcal{X})$ is given by an orthogonal resolution of the identity.

Every von Neumann algebra $\mathfrak{B}$ is a C^*-algebra with identity. A positive map Φ from the algebra $\mathfrak{B}$ is called *normal* if from $X_a \uparrow X$ in $\mathfrak{B}$ it follows that $\Phi(X_a) \uparrow \Phi(X)$.

Corollary 3.1.1 ([150]). *Every normal completely positive map* $\Phi : \mathfrak{B}(\mathcal{H}) \to \mathfrak{B}(\mathcal{H})$ *has the form*

$$\Phi(X) = \sum_{n=1}^{\infty} V_n^* X V_n, \tag{3.4}$$

where the series $\sum_{n=1}^{\infty} V_n^* V_n$ *converges strongly in* $\mathfrak{B}(\mathcal{H})$.

Proof. If Φ is normal then the representation π in formula (3.3) may also be taken normal. It is known that every normal representation of the algebra $\mathfrak{B}(\mathcal{H})$ in the space $\tilde{\mathcal{K}}$ is equivalent to a multiple of the identical representation, that is, $\tilde{\mathcal{K}} = \mathcal{H} \otimes \mathcal{H}_0$ and $\pi(X) = X \otimes I_0$, where $\mathcal{H}_0$ is another Hilbert space and I_0 is the unit operator in $\mathcal{H}_0$ (see, for example, [56] Chap. 9). Let $\{e_j\}$ be an orthonormal basis in $\mathcal{H}_0$, then $V\psi = \sum_{n=1}^{\infty} V_n \psi \otimes e_n$, and (3.3) is transformed into (3.4).

A review of properties of completely positive maps can be found in Størmer's paper in [71]. Ando and Choi [12] considered nonlinear completely positive mappings and established a corresponding generalization of the Stinespring's theorem.

3.1.2 Operations, Dynamical Maps

As a Banach space, the algebra $\mathfrak{B}(\mathcal{H})$ is dual to the space of trace-class operators $\mathfrak{T}(\mathcal{H})$ (see Sect. 1.1.1 in Chap. 1). If Ψ is a linear map in $\mathfrak{T}(\mathcal{H})$, which is positive in the sense that $\Psi(T) \geq 0$ whenever $T \geq 0$, then Ψ is bounded (see, for example, [56] Chap. 2) and hence has the adjoint map $\Phi = \Psi^* : \mathfrak{B}(\mathcal{H}) \to \mathfrak{B}(\mathcal{H})$, which is a positive normal linear map; furthermore, every such Φ is adjoint to some Ψ. If, in addition, $\operatorname{Tr} \Psi(T) \leq \operatorname{Tr} T$ for all $T \in \mathfrak{T}(\mathcal{H}), T \geq 0$, then Ψ is called an *operation* (in the state space). In terms of the adjoint map, this is equivalent to the condition $\Phi(I) \leq I$. The map Φ is also called an operation (in the algebra of observables).

In quantum statistics operations describe changes of states (or of observables) of an open system resulting from evolution or from a macroscopic influence including possible selection according to some criterion (such as measurement outcomes) in the corresponding statistical ensemble. If S is the density operator of the initial state, then the number $\operatorname{Tr} \Psi(S)$ is interpreted as the proportion of the selected representatives of the ensemble, and $\frac{\Psi(S)}{\operatorname{Tr} \Psi(S)}$ as the density operator describing the new state of the selected ensemble. The term "operation" was introduced in the famous work of Haag and Kastler [90] on the foundations of quantum field theory. Based on the concept of

operation, an axiomatic approach to quantum mechanics was developed by
Ludwig, Davies and Lewis, and by other authors (see, for example, [56], [161],
[151], [71], [42]).

Of special importance in dynamical theory are operations transforming
states into states. This is expressed by the property $\operatorname{Tr}\Psi(T) = \operatorname{Tr} T$ for all
$T \in \mathfrak{T}(\mathcal{H})$ or that $\Phi(I) = I$, where $\Phi = \Psi^*$. If, in addition, Φ is completely
positive, then Ψ (or Φ) is called *dynamical map*.[1] From Wigner's theorem
(see Sect. 1.2.1 in Chap. 1) it follows that reversible dynamical maps are of
the form

$$\Psi(T) = UTU^*, \quad \Phi(X) = U^*XU,$$

where U is a unitary operator in $\mathcal{H}$ (if U is antiunitary then Ψ, Φ are antilin-
ear, and we shall not consider this case, which is related to time inversion). In
the general case, dynamical maps describe irreversible evolutions and form a
noncommutative analogue of (transition) Markov maps in probability theory.

An important measure of irreversibility is provided by the *relative entropy*
of the quantum states

$$H(S_1; S_2) = \begin{cases} \operatorname{Tr} S_1\left(\log S_1 - \log S_2\right), & \text{if} \quad \operatorname{supp} S_2 \geq \operatorname{supp} S_1, \\ +\infty & \text{otherwise} \end{cases},$$

where supp$S$ denotes closure of the range of the operator S. Concerning accurate definition and the properties of the relative entropy, see [226], [177].
Most important is *monotonicity* or the generalized H-theorem : for an arbitrary dynamical map Ψ:

$$H(\Psi(S_1); \Psi(S_2)) \leq H(S_1; S_2),$$

which was proved by Lindblad [158], using the fundamental property of *strong
subadditivity* of the quantum entropy, established by Lieb and Ruskai (see
the survey [226]). The proof of monotonicity is by no means simple, and only
recently Lesniewski and Ruskai found a direct argument, that makes no use
of strong subadditivity [155].

According to the Corollary of Sect. 3.1.1, a dynamical map admits the
representation

$$\Psi(T) = \sum_{n=0}^{\infty} V_n T V_n^*, \quad \Psi(X) = \sum_{n=0}^{\infty} V_n^* X V_n,$$

in which $\sum_{n=0}^{\infty} V_n^* V_n = I$. From this it is easy to obtain the

---

[1] The property of complete positivity of the evolution of an open system was
stressed, in particular, in [150], [96],[159].

Corollary 3.1.2. *A map* $\Psi : \mathfrak{T}(\mathcal{H}) \to \mathfrak{T}(\mathcal{H})$ *is a dynamical map if and only if there exist a Hilbert space* $\mathcal{H}_0$, *a state* S_0 *in* $\mathcal{H}_0$ *and a unitary operator* U *in* $\tilde{\mathcal{H}} = \mathcal{H} \otimes \mathcal{H}_0$, *such that*

$$\Psi(S) = \text{Tr}_{\mathcal{H}_0} U(S \otimes S_0)U^*,$$

where $\text{Tr}_{\mathcal{H}_0}$ *is the partial trace over* $\mathcal{H}_0$.

In this way a dynamical map is dilated to a reversible evolution of the composite system consisting of the initial open system and the environment, the very possibility of such a dilation relying upon the property of complete positivity.

3.1.3 Expectations

These are the maps $\mathcal{E} : \mathfrak{B}(\mathcal{H}) \to \mathfrak{B}(\mathcal{H})$, which are idempotent ($\mathcal{E}^2 = \mathcal{E}$) and have unit norm. An expectation $\mathcal{E}$ maps $\mathfrak{B}(\mathcal{H})$ onto a C*-subalgebra $\mathfrak{A} = \{X : X \in \mathfrak{B}(\mathcal{H}), \mathcal{E}(X) = X\}$. If $\mathcal{E}$ is normal, then $\mathfrak{A}$ is a von Neumann algebra. The conditional expectation is *compatible* with the state S, if $\text{Tr}\, S\mathcal{E}(X) = \text{Tr}\, SX$ for $X \in \mathfrak{B}(\mathcal{H})$. Tomiyama showed that a conditional expectation is a positive map and has the property

$$\mathcal{E}(XYZ) = X\mathcal{E}(Y)Z \quad \text{for } Y \in \mathfrak{B}(\mathcal{H}), X, Z \in \mathfrak{A}.$$

This property is included in the original definition given by Umegaki in [220]. In fact every conditional expectation is completely positive.

Takesaki [217] gave a general criterion for the existence of normal conditional expectations in von Neumann algebras. This criterion has the following formulation in the case of $\mathfrak{B}(\mathcal{H})$. Let S be a nondegenerate density operator. Associated with it there is the *modular group of automorphisms of* $\mathfrak{B}(\mathcal{H})$

$$\alpha_t(X) = S^{it} X S^{-it} \quad \text{for } X \in \mathfrak{B}(\mathcal{H}), t \in \mathbb{R} .$$

A conditional expectation $\mathcal{E}$ onto the subalgebra $\mathfrak{A} \subset \mathfrak{B}(\mathcal{H})$ compatible with the state S exists if and only if $\mathfrak{A}$ is invariant under the α_t. In particular, this occurs if $[S, X] = 0$ for all $X \in \mathfrak{A}$.

An example of a normal conditional expectation (averaging over a subsystem of a composite system) was given in Sect. 1.3.1, Chap. 1. Another important example is the following: let $\{E_n\}$ be an orthogonal resolution of the identity in $\mathcal{H}$. Then

$$E(X) := \sum_{n=1}^{\infty} E_n X E_n \tag{3.5}$$

is a normal conditional expectation onto the subalgebra $\mathfrak{A}$ of operators of form (3.5), compatible with any state whose density operator belongs to $\mathfrak{A}$.

The algebra $\mathfrak{A}$ can be characterized as $\mathfrak{A} = \{E_n; n = 1, 2, \ldots\}'$, where $\mathfrak{M}'$ denotes the *commutant* of the subset $\mathfrak{M} \subset \mathfrak{B}(\mathcal{H})$, i.e. the set of all bounded operators commuting with all operators in $\mathfrak{M}$.

A review of results concerning expectations in von Neumann algebras can be found in the papers by Cecchini in [193] and Petz in [195].

3.2 Quantum Channel Capacities

3.2.1 The Notion of Channel

Information theory deals with problems in a sense opposite to nonequilibrium statistical mechanics: given an irreversible transformation, such as noisy channel, make it useful for almost perfect (reversible) data transmission. The coding theorem establishes a fundamental characteristic of the channel – its capacity – which describes the exponential rate of the growth of the maximal size of data array that can be thus transmitted by n independent uses of the channel [54], as $n \to \infty$.

The information is encoded into states of physical systems. Let $\mathfrak{A}$ (resp. $\mathfrak{B}$) be a C*-algebra with unit, which we shall call the *input* (resp. *output*) algebra. We denote by $\mathfrak{S}(\mathfrak{A})$ the set of states on $\mathfrak{A}$. Usually $\mathfrak{A}$ is a von Neumann algebra and $\mathfrak{S}(\mathfrak{A})$ is the set of normal states. However, in many important applications $\mathfrak{A}$ is finite-dimensional (which we also assume for simplicity below), and the topological complications become irrelevant. A *channel* is an affine map Φ from $\mathfrak{S}(\mathfrak{A})$ to $\mathfrak{S}(\mathfrak{B})$, the linear extension of which (still denoted Φ) has completely positive adjoint Φ^*. The map Φ is in general non-invertible, thus representing an irreversible (noisy) evolution. In particular, if $\mathfrak{A}$ (resp. $\mathfrak{B}$) is Abelian, Φ may describe encoding (resp. decoding) of classical information; if both are Abelian, Φ is a classical channel. Let us illustrate these notions by simple examples.

Classical (c-c) channels Let $\mathcal{X}$, $\mathcal{Y}$ be finite sets (alphabets), $\mathfrak{A} = C(\mathcal{X})$, $\mathfrak{B} = C(\mathcal{Y})$ be the algebras of all complex functions on $\mathcal{X}$, $\mathcal{Y}$. Then any channel is described by a transition probability from $\mathcal{X}$ to $\mathcal{Y}$.

Quantum (q-q) channels Let $\mathcal{H}$, $\mathcal{K}$ be finite-dimensional Hilbert spaces, and let $\mathcal{A} = \mathfrak{B}(\mathcal{H})$, $\mathcal{B} = \mathfrak{B}(\mathcal{K})$ be the algebras of all linear operators in $\mathcal{H}$, $\mathcal{K}$. Then any channel from $\mathfrak{S}(\mathfrak{A})$ to $\mathfrak{S}(\mathfrak{B})$ is given by the Stinespring-Kraus representation

$$\Phi(S) = \sum_j V_j S V_j^*, \quad S \in \mathfrak{S}(\mathfrak{A}),$$

where $\sum_j V_j^* V_j = I_{\mathcal{H}}$ (the unit operator in $\mathcal{H}$). Particularly important classes of channels in quantum optics are *qubit channels* for which $\dim\mathcal{H} = \dim\mathcal{K} = 2$ [32], and *Gaussian channels* over the algebras of canonical commutation relations (which require infinite dimensional Hilbert spaces) [134], [135].

C-q channels Let $\mathfrak{A} = C(\mathfrak{X})$, $\mathfrak{B} = \mathfrak{B}(\mathfrak{K})$, then a channel from $\mathfrak{S}(\mathfrak{A})$ to $\mathfrak{S}(\mathfrak{B})$ is given by the relation

$$\Phi(P) = \sum_{x \in \mathfrak{X}} p_x S_x,$$

where $P = \{p_x\}$ is a probability distribution (classical state), and S_x are fixed density operators in $\mathfrak{K}$ (signal states).

Q-c channels Let $\mathfrak{A} = \mathfrak{B}(\mathfrak{H})$, $\mathfrak{B} = C(\mathcal{Y})$, then

$$\Phi(S) = \{\text{Tr} S M_y\}_{y \in \mathcal{Y}},$$

where $\{M_y\}_{y \in \mathcal{Y}}$ is a resolution of the identity in $\mathfrak{H}$, i. e. a collection of hermitean operators, satisfying $M_y \geq 0, \sum M_y = I_{\mathfrak{H}}$. In particular, an orthogonal resolution of the identity, for which $M_y^2 = M_y$, corresponds to a von Neumann measurement with the outcomes $y \in \mathcal{Y}$.

3.2.2 The Variety of Classical Capacities

The process of information transmission is represented by the diagram

$$\mathfrak{S}(\mathfrak{A}) \quad \overset{\mathfrak{E}}{\to} \overset{\Phi}{\to} \overset{\mathfrak{D}}{\to} \quad \mathfrak{S}(\mathfrak{B}). \tag{3.6}$$

Here the (noisy) channel Φ is given and the purpose of the information systems design is to choose $\mathfrak{E}$ and $\mathfrak{D}$ so as to minimize the distortion of the input signal due to the channel noise. Typically one considers an asymptotic situation, where a sequence of channels is given, such as $\Phi^{\otimes n} = \overset{n}{\overbrace{\Phi \otimes \cdots \otimes \Phi}}$, with $n \to \infty$, Φ fixed, and the question is how quickly does the information content which can be transmitted asymptotically precisely through $\Phi^{\otimes n}$ grow. The last model is called a *memoryless channel*. More generally, one can consider an (open) dynamical system with ergodic behavior and the limit where the observation time goes to infinity.

To define the classical capacity of the channel Φ we take $\mathfrak{A}$, $\mathfrak{B}$ in (3.6) Abelian, namely $\mathfrak{A}_m = \mathfrak{B}_m = C(\{0, 1\}^{\otimes m})$, and denote by

$$\lambda_c(m, n) = \inf_{\mathfrak{E}_n^m, \mathfrak{D}_n^m} \|\text{Id}_2^{\otimes m} - \mathfrak{D}_n^m \circ \Phi^{\otimes n} \circ \mathfrak{E}_n^m\| \tag{3.7}$$

the quantity which measures the deviation of the composition: encoding – channel – decoding from the ideal (identity) channel $\text{Id}_2^{\otimes m} : \mathfrak{S}(\mathfrak{A}_m) \to \mathfrak{S}(\mathfrak{B}_m)$. Here the infimum is taken over all possible encodings and decodings. We consider the limit $n \to \infty, m/n = R$ fixed. The *classical capacity* of Φ is defined as

$$C(\Phi) = \sup\{R : \lim_{n \to \infty} \lambda_c(nR, n) = 0\}. \tag{3.8}$$

To define the quantum capacity we take $\mathfrak{A}_m = \mathfrak{B}_m = \mathfrak{B}(\mathcal{H}_2^{\otimes m})$, where $\mathcal{H}_2$ is the two-dimensional Hilbert space (qubit), and consider the quantity $\lambda_q(m,n)$ given by the formula similar to (3.7), where the infimum is taken over the corresponding classes of encodings and decodings. The *quantum capacity* $Q(\Phi)$ is then given by the formula of the type (3.8) with $\lambda_q(m,n)$ replacing $\lambda_c(m,n)$.

Let $\mathcal{H}$ be a Hilbert space and $\Phi : \mathfrak{S}(\mathcal{H}) \to \mathfrak{S}(\mathcal{H})$ – a quantum channel. The *coding theorem* for the c-q channels (see Sect. 2.2.3 of Chap. 2) implies that

$$C(\Phi) = \lim_{n\to\infty} \frac{1}{n}\bar{C}(\Phi^{\otimes n}), \qquad (3.9)$$

where

$$\bar{C}(\Phi) = \max\left\{ H\left(\sum_x p_x \Phi(S_x)\right) - \sum_x p_x H(\Phi(S_x)) \right\}, \qquad (3.10)$$

$H(S) = -\mathrm{Tr}S \log S$ is the entropy, and the maximum is taken over all probability distributions $\{p_x\}$ and collections of density operators $\{S_x\}$ in $\mathcal{H}$. It is still not known whether the quantity $\bar{C}(\Phi)$ is additive, i.e.

$$\bar{C}(\Phi^{\otimes n}) \stackrel{?}{=} n\bar{C}(\Phi)$$

for arbitrary quantum channel Φ. This holds for classical channels, c-q, q-c channels and for ideal quantum channels. Further partial results can be found in [11].

Additivity of $\bar{C}(\Phi)$ would have an important physical consequence – it would mean that using entangled input states does not increase the classical capacity of a quantum channel. On the other hand, it is known that using entangled measurements at the output of a quantum channel can indeed increase the capacity (see Sect. 2.2.3 of Chap. 2). More generally, following Bennett and Shor [32], we can define four classical capacities $C_{1,1}, C_{1,\infty}, C_{\infty,1}, C_{\infty,\infty}$ of a channel Φ, where the first (second) index refers to encodings (decodings), ∞ means using arbitrary encodings (decodings), 1 means restriction to encodings with unentangled outputs (decodings with unentangled inputs) in the definition (3.7). Then $C_{\infty,\infty} = C(\Phi), C_{1,\infty} = \bar{C}(\Phi)$, and $C_{1,1}$ is the "one-shot" capacity, which is the maximal Shannon information, accessible by using unentangled inputs and outputs. The relations between these four capacities is shown on the following diagram:

$$\begin{array}{ccc} C_{\infty,1} & \stackrel{<}{\neq} & C_{\infty,\infty} \\ \| & ? & \\ C_{1,1} & \stackrel{<}{\neq} & C_{1,\infty} \end{array} \qquad (3.11)$$

where the symbol $\stackrel{<}{\neq}$ means "always less than or equal to and sometimes strictly less". The equality $C_{1,1} = C_{\infty,1}$ follows by argument similar to Proposition

1 in [133]; the upper inequality $C_{\infty,1} \overset{<}{\neq} C_{\infty,\infty}$ follows from this and from the lower inequality, which expresses the strict superadditivity with respect to decodings.

Example 3.2.1. Consider the configuration of the three equiangular vectors $|\psi_k\rangle$, $k = 1, 2, 3$, in a 2-dimensional space, as shown on Fig. 2.1 in Sect. 2.1.3, and the corresponding c-q channel

$$\Phi(P) = \sum_{k=1}^{3} p_k |\psi_k\rangle\langle\psi_k| \equiv \bar{S}_P.$$

This channel was introduced in [97] to show that for the uniform P the maximal Shannon information accessible through arbitrary von Neumann measurements (≈ 0.429 bits) is *strictly less* than the maximum over all decodings given by arbitrary resolutions of the identity (≈ 0.585 bits). The quantity $C_{1,1} \approx 0.645$ bits is attained for the distribution $P = (\frac{1}{2}, \frac{1}{2}, 0)$ and the measurement corresponding to the orthonormal basis, optimal for discrimination between the two equiprobable states $|\psi_k\rangle$, $k = 1, 2$. On the other hand, it is easy to see that $C_{\infty,\infty}(= C_{1,\infty}) = \max_P H(\bar{S}_P) = 1$ with the maximum achieved on the uniform distribution. As follows from (2.20) of Chap. 2, the value $C(\Phi) = 1$ bit is the absolute maximum for the capacity of a qubit channel. This gives an example of channel with nonorthogonal states which nevertheless has this maximal capacity.

3.2.3 Quantum Mutual Information

Let S be an input state for the channel Φ. There are three important entropy quantities related to the pair (S, Φ), namely, the entropy of the input state $H(S)$, the entropy of the output state $H(\Phi(S))$, and the entropy exchange $H(S, \Phi)$. While the definition and the meaning of the first two entropies is clear, the third quantity is somewhat more sophisticated. To define it, one introduces the *reference system*, described by the Hilbert space $\mathcal{H}_R$, isomorphic to the Hilbert space $\mathcal{H}_Q = \mathcal{H}$ of the initial system. Then according to [22], there exists a *purification* of the state S, i.e. a unit vector $|\psi\rangle \in \mathcal{H}_Q \otimes \mathcal{H}_R$ such that

$$S = \operatorname{Tr}_R |\psi\rangle\langle\psi|.$$

The *entropy exchange* is then defined as

$$H(S, \Phi) = H\big((\Phi \otimes \operatorname{Id}_R)(|\psi\rangle\langle\psi|)\big), \tag{3.12}$$

that is, as the entropy of the output state of the dilated channel $(\Phi \otimes \operatorname{Id}_R)$ applied to the input which is purification of the state S. It is easily shown to be independent on the choice of a purification.

From these three entropies one can construct several information quantities. In analogy with classical information theory, one can define *quantum mutual information* between the reference system R (which mirrors the input Q) and the output of the system Q' [5] as

$$I(S, \Phi) = H(S'_R) + H(S'_Q) - H(S'_{RQ}) = H(S) + H(\Phi(S)) - H(S, \Phi). \tag{3.13}$$

The quantity $I(S, \Phi)$ has a number of "natural" properties, in particular, positivity, concavity with respect to the input state S and additivity for product channels which are established with the help of strong subadditivity [5]. Moreover, the maximum of $I(S, \Phi)$ with respect to S was proved recently to be equal to the *entanglement-assisted classical capacity* $C_{ea}(\Phi)$ of the channel [33]. This means that the input and the output of the channel Φ are allowed to share a common entangled state, the input part of which is used for encoding classical information and then sent to the output via the channel Φ. It was shown that this maximum is additive for the product channels, the one-shot expression thus giving the full (asymptotic) capacity.

In general, $C_{ea}(\Phi) \geq C(\Phi)$, and for the ideal channel $C_{ea}(\mathrm{Id}) = 2C(\mathrm{Id})$. For non-ideal channels the ratio $C_{ea}(\Phi) : C(\Phi)$ in general can increase indefinitely with the decrease of signal-to noise ratio [33], [135]. Thus, although sole sharing entanglement between input and output cannot be used for classical information transmission, it can play a role of catalyzer, enhancing the information transmission through a genuine communication channel between $\mathfrak{A}$ and $\mathfrak{B}$.

3.2.4 The Quantum Capacity

An important component of $I(S, \Phi)$ is the *coherent information*

$$J(S, \Phi) = H(\Phi(S)) - H(S, \Phi), \tag{3.14}$$

the maximum of which has been conjectured to be the (one-shot) quantum capacity of the channel Φ [22]. Its properties are not so nice. It can be negative, it is difficult to maximize, and its maximum was shown to be strictly superadditive for certain product channels, hence the conjectured full quantum capacity can be greater than the one-shot expression, in contrast to the case of the entanglement-assisted classical capacity.

By using quantum generalization of the Fano inequality, Barnum, Nielsen and Schumacher [22] (see also [23]) were able to prove the converse coding theorem for the quantum capacity $Q(\Phi)$, namely

$$Q(\Phi) \leq \limsup_{n \to \infty} \frac{1}{n} \max_S J(S, \Phi^{\otimes n}). \tag{3.15}$$

The right-hand side of this relation is not easy to evaluate even for the most simple binary channels. A useful and analytically more tractable bound is given by the Werner inequality

$$Q(\Phi) \leq \log \|\Phi \circ \Theta\|_{cb},$$

where Θ is a transposition map and $\| \cdot \|_{cb}$ denotes the norm of complete boundedness (see [135], where this bound as well as the entropy and information quantities are computed and compared for quantum Gaussian channels).

One of the great achievements of the quantum information theory is discovery of quantum error correcting codes allowing error-free transmission of quantum information through noisy channels [32], [212], and one of the challenges is the problem of quantum capacity: the conjecture is that in fact it is equal to the right hand side of (3.15).

3.3 Quantum Dynamical Semigroups

3.3.1 Definition and Examples

A dynamical semigroup is the noncommutative generalization of a semigroup of transition operators in the theory of Markov random processes. It describes irreversible evolution of quantum system determined only by the present state of the system, without memory of the past. In quantum statistical mechanics dynamical semigroups arise from considerations of weak or singular coupling limits for open quantum systems interacting with the environment [209]. These semigroups satisfy differential equations that are noncommutative generalizations of the Fokker-Planck or Chapman-Kolmogorov equations. There are two equivalent ways of defining a dynamical semigroup - in the state space and in the algebra of observables of the system.

A *quantum dynamical semigroup* in the state space is a family of dynamical maps $\{\Psi_t; t \in \mathbb{R}_+\}$ in the Banach space of trace class operators $\mathfrak{T}(\mathcal{H})$ such that

1. $\Psi_t \cdot \Psi_s = \Psi_{t+s}$ for $t, s \in \mathbb{R}_+$;
2. $\Psi_0 = \mathrm{Id}$ (the identity map);
3. $\{\Psi_t\}$ is *strongly continuous*, i.e. $\lim_{t \to 0} \|\Psi_t(T) - T\|_1 = 0$ for arbitrary $T \in \mathfrak{T}(\mathcal{H})$.

From the general theory of semigroups in a Banach space (see, for example, [40] Chap. 3) it follows that there exists a densely defined infinitesimal generator

$$\mathcal{K}(T) = \lim_{t \to 0} \frac{\Psi_t(T) - T}{t}.$$

If Ψ_t is continuous in norm, i.e. $\lim_{t\to 0}\|\Psi_t - \mathrm{Id}\| = 0$, then $\mathcal{K}$ is an everywhere defined bounded map on $\mathfrak{T}(\mathcal{H})$. If S_0 is an initial state then the function $t \to S_t := \Psi_t(S_0)$ satisfies the quantum Markov master equation

$$\frac{dS_t}{dt} = \mathcal{K}(S_t), \qquad (3.16)$$

which is the noncommutative analogue of the forward Chapman-Kolmogorov equation. In physical applications dynamical semigroups arise as general solutions of the Cauchy problem for Markov master equations.

A dynamical semigroup in the algebra of observables $\{\Phi_t; t \in \mathbb{R}_+\}$ is a semigroup of dynamical maps of the algebra $\mathfrak{B}(\mathcal{H})$, such that $\Phi_0 = \mathrm{Id}$ and $\mathrm{w}^* - \lim_{t\to 0} \Phi_t(X) = X$ for arbitrary $X \in \mathfrak{B}(\mathcal{H})$. Every such semigroup $\{\Phi_t\}$ is adjoint to a semigroup $\{\Psi_t\}$ in $\mathfrak{T}(\mathcal{H})$ and vice versa. The semigroup $\{\Psi_t\}$ is norm continuous if and only if $\{\Phi_t\}$ is such. Let $\mathcal{L} = \mathcal{K}^*$ be the generator of $\{\Phi_t\}$. The family $X_t = \Phi_t(X_0)$ satisfies the backward Markov master equation

$$\frac{dX_t}{dt} = \mathcal{L}(X_t). \qquad (3.17)$$

In the norm continuous case the forward and the backward equations (3.16), (3.17) are equivalent and have unique solutions given by the corresponding dynamical semigroups. This is no longer necessarily true for unbounded generators (see Sect. 3.3.3).

Example 3.3.1 ([149]). Let G be a separable locally compact group, let $g \to V_g$ be a continuous unitary representation of G in $\mathcal{H}$ and let $\{\mu_t; t \in \mathbb{R}_+\}$ be a continuous convolution semigroup of probability measures on G (see, for example, [85]). Then the equations

$$\Psi_t(S) = \int_G V_g S V_g^* \mu_t(dg), \quad \Phi_t(X) = \int_G V_g^* X V_g \mu_t(dg)$$

define quantum dynamical semigroups in $\mathfrak{T}(\mathcal{H})$ and in $\mathfrak{B}(\mathcal{H})$ respectively.

In particular, let A be a Hermitian operator in $\mathcal{H}$. Then the expression

$$\Psi_t(S) = \frac{1}{\sqrt{2\pi t}} \int_{-\infty}^{\infty} e^{-\frac{x^2}{2t}} e^{-iAx} S e^{iAx} dx,$$

corresponding to the Gaussian convolution semigroup on $\mathbb{R}$, gives a dynamical semigroup with the infinitesimal generator

$$\mathcal{K}(S) = ASA - A^2 \circ S. \qquad (3.18)$$

If U is a unitary operator and $\lambda > 0$, then

$$\Psi_t(S) = \sum_{n=0}^{\infty} \frac{(\lambda t)^n}{n!} e^{-\lambda t} U^n S U^{*n}$$

is a dynamical semigroup, corresponding to the Poisson convolution semigroup on $\mathbb{Z}$, with infinitesimal generator

$$\mathcal{K}(S) = \lambda(USU^* - S). \tag{3.19}$$

In both cases we have a stochastic representation

$$\Psi_t(S) = \mathbf{E}(e^{-iA\xi_t} S e^{iA\xi_t}),$$

where $\{\xi_t; t \geq 0\}$ is the standard Wiener process in the first case, and the Poisson process with intensity λ in the second case, with $U = e^{iA}$. These are the examples of irreversible quantum evolutions due to interactions with classical (white Gaussian and Poisson shot) noises.

The concept of dynamical semigroup was proposed by Kossakowski [149] (see also [56]), however without the crucial condition of complete positivity which was introduced later by Lindblad [159]. Many physical examples are included in the general scheme of quasifree dynamical semigroups which are the quantum analogue of Gaussian Markov semigroups. In the case of the CCR such semigroups are characterized by the condition that they transform Gaussian states into Gaussian states (see Sect. 1.2.4 in Chap. 1). In statistical mechanics they describe the irreversible dynamics of open Bose or Fermi systems with quadratic interaction (see the reviews [178], [10]).

3.3.2 The Generator

The requirement of complete positivity imposes nontrivial restrictions on the generator of the semigroup. A description of generators of norm-continuous quantum dynamical semigroups was obtained by Lindblad and, independently, in the case $\dim \mathcal{H} < \infty$ by Gorini, Kossakowski and Sudarshan (see [159], [69]).

Theorem 3.3.1 ([159]). *A bounded map $\mathcal{K}$ on the space $\mathfrak{T}(\mathcal{H})$ is the generator of a norm continuous quantum dynamical semigroup if and only if*

$$\mathcal{K}(S) = -i[H, S] + \sum_{n=0}^{\infty} (L_j S L_j^* - L_j^* L_j \circ S), \tag{3.20}$$

where $H, L_j \in \mathfrak{B}(\mathcal{H})$, $H = H^*$ and the series $\sum_{j=1}^{\infty} L_j^* L_j$ converges strongly.

The first term in (3.20) corresponds to a reversible evolution with Hamiltonian H, while the second represents the dissipative part. The operator

$\sum_{j=1}^{\infty} L_j^* L_j$ is related to the rate of the dissipation. Passing over to the formulation in the algebra of observables, for the generator $\mathcal{L} = \mathcal{K}^*$ of the semigroup $\Phi_i = \Psi_t^*$ we have

$$\mathcal{L}(X) = i[H, X] + \sum_{j=1}^{\infty} (L_j^* X L_j - L_j^* L_j \circ X). \tag{3.21}$$

The proof of the theorem rests on the following two facts.

Proposition 3.3.1. *Let $\mathcal{L}$ be a bounded map from $\mathfrak{B}(\mathcal{H})$ to itself, such that $\mathcal{L}[I] = 0$. The following statements are equivalent:*

1. *$\exp t\mathcal{L}$ is completely positive for all $t \in \mathbb{R}_+$;*
2. *$\mathcal{L}[X^*] = \mathcal{L}[X]^*$ and $\mathcal{L}$ is conditionally completely positive, that is, from $\sum_j X_j \psi_j = 0$ it follows that*

$$\sum_{j,k} \langle \psi_j | \mathcal{L}[X_j^* X_k] \psi_k \rangle \geq 0.$$

The proof of this Proposition, which is a noncommutative analog of Schoenberg's theorem in the theory of positive definite functions (see, for example, [186], also Sect. 4.2.4 in Chap. 4), can be found in [69]. The implication 1) $\Longrightarrow$ 2) follows by differentiation of $\exp t\mathcal{L}$ at $t = 0$.

Theorem 3.3.2. *A bounded map $\mathcal{L}$ of $\mathfrak{B}(\mathcal{H})$ is conditionally completely positive if and only if*

$$\mathcal{L}[X] = \Phi[X] - K^* X - XK, \tag{3.22}$$

where Φ is a completely positive map and $K \in \mathfrak{B}(\mathcal{H})$.

There is an elementary proof which, moreover, can be generalized to the case of unbounded generators [124]. Let us fix a unit vector $\psi_0 \in \mathcal{H}$ and define operator K by

$$\langle \psi | K \phi \rangle = -\langle \psi_0 | \mathcal{L}[|\psi_0\rangle\langle\psi|] \phi \rangle + \frac{1}{2} \langle \psi_0 | \mathcal{L}[|\psi_0\rangle\langle\psi_0|] \psi_0 \rangle \langle \psi | \phi \rangle.$$

Defining the map $\Phi[X]$ by $\Phi[X] = \mathcal{L}[X] - K^* X - XK$, we have

$$\sum_{j,k=1}^{n} \langle \psi_j | \Phi[X_j^* X_k] \psi_k \rangle = \sum_{j,k=0}^{n} \langle \psi_j | \mathcal{L}[X_j^* X_k] \psi_k \rangle$$

for any finite collections $\{\psi_j\}_{j=1,\dots,n} \in \mathcal{H}$; $\{X_j\}_{j=1,\dots,n} \in \mathfrak{B}(\mathcal{H})$, where $X_0 = -\sum_{j=1}^{n} X_j |\psi_j\rangle\langle\psi_0|$. Since $\sum_{j=0}^{n} X_j \psi_j = 0$, and $\mathcal{L}$ is conditionally completely positive, $\Phi(X)$ is completely positive.

The representation (3.21) is then obtained from the formula (3.3) for a normal completely positive map, by taking into account the normalization condition $\mathcal{L}(I) = 0$.

The proof of the formula (3.22) in the case of arbitrary von Neumann algebra $\mathfrak{A}$ can be related to the cohomology of $\mathfrak{A}$ [52]. A GNS type construction associates with the conditionally completely positive map $\mathcal{L}$ a linear map B from the algebra $\mathfrak{A}$ to the space of bounded operators from $\mathcal{H}$ into another Hilbert space $\mathcal{K}$ such that

$$\mathcal{L}[X^*Y] - X^*\mathcal{L}[Y] - \mathcal{L}[X^*]Y = B[X]^*B[Y].$$

Furthermore the map B is a *cocycle* for a representation π of the algebra $\mathfrak{A}$ into $\mathfrak{A}$, that is, satisfies the equation

$$B[XY] = \pi[X]B[Y] + B[X]Y; \quad X, Y \in \mathfrak{A}.$$

The main point of the proof is the fact that every bounded cocycle is trivial, that is, it has the form $B[X] = \pi[X]R - RX$, where R is a bounded operator from $\mathcal{H}$ to $\mathcal{K}$.

A physical interpretation for the standard representation can be seen from the Dyson expansion of the solution of the forward Markov master equation

$$\Psi_t[S] = \hat{\Psi}_t[S] + \sum_{n=1}^{\infty} \int \cdots \int_{0 \leq t_1 \leq \cdots \leq t_n \leq t} \hat{\Psi}_{t_1}[\Psi[\hat{\Psi}_{t_2-t_1} \cdots \Psi[\hat{\Psi}_{t-t_n}[S]] \cdots]] dt_1 \ldots dt_n$$

described as sequence of "spontaneous jumps" of the magnitude $\Psi[S] = \sum_j LSL_j^*$ occurring at times $t_1 \leq \cdots \leq t_n$ on the background of the "relaxing evolution" given by the semigroup $\hat{\Psi}_t[S] = e^{-Kt}Se^{-K^*t}$.

3.3.3 Unbounded Generators

Dynamical semigroups without the property of norm continuity, while being interesting from both physical and mathematical points of view, pose difficult problems. The generator of such a semigroup is unbounded, and its domain can be quite complicated and difficult to describe. Therefore it is not clear how to generalize the property of conditional complete positivity and characterize the generator of the semigroup. Already the problem of constructing a dynamical semigroup corresponding to a formal expression of the type (3.21), where H, L_j are unbounded operators, is nontrivial. Davies [57] gave conditions under which with the formal expression (3.20) is associated a strongly continuous semigroup of completely positive maps $\{\Phi_t; t \in \mathbb{R}_+\}$, such that

$$\Phi_t[I] \leq I,$$

which is the analogue of the Feller minimal solution in the classical theory of Markov processes.

The more recent probabilistic approach [124], [50] shifts the accent from the dynamical semigroup to the Markov master equation it satisfies. Let K, L_j be operators defined on a dense domain $\mathcal{D} \subset \mathcal{H}$ and satisfying the *conservativity condition*

$$\sum_j \|L_j \psi\|^2 = 2\Re\langle \psi | K\psi \rangle, \quad \psi \in \mathcal{D}. \qquad (3.23)$$

This implies, in particular, that the operator K is *accretive*: $\Re\langle \psi | K\psi \rangle \geq 0, \psi \in \mathcal{D}$. Introducing the *form-generator*

$$\mathcal{L}(\psi; X; \phi) = \sum_j \langle L_j \psi | X L_j \phi \rangle - \langle K\psi | X\phi \rangle - \langle \psi | X K\phi \rangle, \quad \phi, \psi \in \mathcal{D}, \quad (3.24)$$

consider the differential equation

$$\frac{d}{dt}\langle \psi | \Phi_t[X]\psi \rangle = \mathcal{L}(\psi; \Phi_t[X]; \phi), \qquad (3.25)$$

which we call *backward Markov master equation*. By a solution of this equation we mean a family $\Phi_t, t \geq 0$, of normal completely positive maps, uniformly bounded and w^*-continuous in t and satisfying (3.25) for all X and $\phi, \psi \in \mathcal{D}$.

Theorem 3.3.3. *Let K be maximal accretive and $\mathcal{D}$ be an invariant domain of the semigroup $e^{-Kt}, t \geq 0$. Then there exists the minimal solution Φ_t^∞ of the backward equation (3.25), such that for any other solution Φ_t the difference $\Phi_t - \Phi_t^\infty$ is completely positive; moreover, Φ_t^∞ is a semigroup.*

If the minimal semigroup is *unital* in the sense that $\Phi_t^\infty[I] = I$, then it is the unique solution of (3.25).

Under additional assumption that operators L_j^*, K^* are defined on a dense domain $\mathcal{D}^*$ and

$$\sum_j \|L_j^* \psi\|^2 < \infty, \quad \psi \in \mathcal{D}^*,$$

we can consider also the *forward Markov master equation* for the semigroup Ψ_t in the state space:

$$\frac{d}{dt}\langle \psi | \Psi_t[S]\phi \rangle = \qquad (3.26)$$

$$= \sum_j \langle L_j^* \phi | \Psi_t[S] L_j^* \psi \rangle - \langle K^* \phi | \Psi_t[S]\psi \rangle - \langle \phi | \Psi_t[S] K^* \psi \rangle, \quad \phi, \psi \in \mathcal{D}^*.$$

Theorem 3.3.4. *Let K^* be maximal accretive and let $\mathcal{D}^*$ be an invariant domain of the semigroup $e^{-K^*t}, t \geq 0$. Then there exists the minimal solution Ψ_t^∞ of the backward equation* (3.26), *and* $(\Psi_t^\infty)^* = \Phi_t^\infty)$.

Example 3.3.2. Let $\mathcal{H} = l^2$ be the Hilbert space of square summable sequences $\{\psi_n; n \geq 0\}$ and let N, W be the "number" and "shift" operators defined by the relations

$$(N\psi)_n = n\psi_n, (W\psi)_n = \psi_{n-1}$$

on the dense domain $\mathcal{D} = \mathcal{D}^*$ of sequences with finite number of nonzero coefficients ψ_n. Denoting $S_t = \Psi_t[S]$, consider the forward equation

$$\frac{d}{dt}\langle\phi|S_t\psi\rangle = \langle L(N)^*W^*\phi|S_t L(N)^*W^*\psi\rangle$$
$$-\langle K(N)^*\phi|S_t\psi\rangle - \langle\phi|S_t K(N)^*\psi\rangle; \phi,\psi \in \mathcal{D},$$

where $L(n), K(n)$ are complex-valued functions satisfying the conservativity condition $|L(n)|^2 = 2\Re K(n) \equiv \lambda_n$. The diagonal elements $p_n(t); n = 0, 1, \ldots$, of the operator S_t satisfy the pure birth equation

$$\frac{dp_n(t)}{dt} = -\lambda_n p_n(t) + \lambda_{n-1}p_{n-1}(t),$$

whose unique solution satisfies $\sum\limits_{n=0}^{\infty} p_n(t) = 1$ if and only if the following Feller condition holds:

$$\sum_{n=0}^{\infty}\lambda_n^{-1} = \infty.$$

This means that the total mean time the system spends in its phase space is infinite. One easily sees that this is necessary and sufficient condition for the minimal solution Φ_t^∞ of the corresponding backward equation to be unital. The situation is similar to that for Kolmogorov-Feller differential equations in the theory of Markov processes. If Φ_t^∞ is not unital, then there is a positive probability of "explosion" i.e. exit from the phase space during finite time. Solution of the backward equation is then non-unique, and additional "boundary conditions" are required to specify the solution.

Davies gave sufficient conditions for conservativity, suitable for a class of models of quantum diffusion. A necessary and sufficient condition, which is a noncommutative analog of the classical general Feller's criterion, is that the equation

$$\mathcal{L}(\psi; X; \phi) + \langle\phi|X\psi\rangle = 0; \quad \phi,\psi \in \mathcal{D},$$

should have no solutions X satisfying $0 \leq X \leq I$ [50]. There is also more practical general sufficient condition, the essential ingredient of which is the property of *hyperdissipativity*: there exists a strictly positive operator A in $\mathcal{H}$ such that

$$\sum_j \|A^{1/2} L_j \psi\|^2 - 2\Re\langle A\psi | K\psi \rangle \leq c\|A^{1/2}\psi\|^2, \quad \psi \in \mathcal{D}. \tag{3.27}$$

Note that the left hand side can be regarded as a definition of the value $\mathcal{L}(\psi; A; \phi)$ for the unbounded operator A. This property, introduced in [126], generalizes and improves the condition of [50] corresponding to $A = \sum_j L_j^* L_j$. It prevents solutions of the Markov master equation from too fast approach to "boundary", the operator A playing the role of a noncommutative "Liapunov function".

Going back to the problem of standard representation, we can make the following remarks. The fact that a form-generator has the standard representation (3.24) implies the possibility of decomposing the generator $\mathcal{K}$ into completely positive and relaxing parts only on the subspace

$$\mathfrak{D} = \{ S : S = |\psi\rangle\langle\phi|, \quad \phi, \psi \in \mathcal{D} \}, \tag{3.28}$$

which need not be a core for $\mathcal{K}$. If explosion occurs, these two parts need not be separately extendable onto a core for $\mathcal{K}$. On the other hand, generators of different dynamical semigroups restricted to (3.28) can give rise to one and the same standard expression (3.24). One may formalize the notion of standard representation by saying that a dynamical semigroup is *standard* if it can be constructed as the minimal semigroup for some Markov master equation, that is by a completely positive perturbation of a relaxing semigroup. In [125] a possible noncommutative extension of "boundary conditions" for conservative form-generator was proposed as very singular completely positive perturbations vanishing on the dense domain (3.28). By using such a perturbation the author gave a construction of non-standard dynamical semigroup on $\mathfrak{B}(\mathcal{H})$ [125], [130].

3.3.4 Covariant Evolutions

Let $g \to V_g$ be a representation in $\mathcal{H}$ of a group G, describing symmetries of a quantum open system. The dynamical semigroup $\{\Psi_t; t \in \mathbb{R}_+\}$ is said to be *covariant* if

$$\Psi_t[V_g S V_g^*] = V_g \Psi_t[S] V_g^*$$

for all $S \in \mathfrak{S}(\mathcal{H})$, $g \in G$, $t \in \mathbb{R}_+$.

A complete classification of generators of dynamical semigroups covariant with respect to rotational symmetries was obtained in the finite-dimensional case in [84], [14].

Example 3.3.3. Consider the evolution (3.16) of an open system with spin 1/2, where $\dim \mathcal{H} = 2$ (see Sect. 1.1.6 of Chap. 1), covariant with respect to the action of the group $SO(2)$ which corresponds to axial symmetry. The representation has the form $\Phi \to e^{i\psi\sigma_3}$, where $\Phi \in [0, 2\pi]$. The general form of generator of a covariant dynamical semigroup is

$$\mathcal{L}[S] = -i[H, S] + \sum_{j=-1}^{i} c_j(L_j S L_j^* - L_j^* L_j \circ S), \qquad (3.29)$$

where $c_j \geq 0$, $H = \frac{1}{2}\omega_0\sigma_3$, $L_{-1} = \frac{1}{\sqrt{2}}(\sigma_1 - i\sigma_2)$, $L_0 = \sigma_3$, and $L_1 = \frac{1}{\sqrt{2}}(\sigma_1 + i\sigma_2)$. Let $S_t = S(\mathfrak{a}_t)$, where $\mathfrak{a}_t$ is the real 3-dimensional vector representing the state. Then we have the Bloch equation for the vector $\mathfrak{a}_t$

$$\frac{d\mathfrak{a}_t}{dt} = \begin{bmatrix} -T_\perp^{-1} & \omega_0 & 0 \\ -\omega & -T_\perp^1 & 0 \\ 0 & 0 & T^{-1} \end{bmatrix} \mathfrak{a}_t + \begin{bmatrix} 0 \\ 0 \\ c_\infty/T_0 \end{bmatrix}, \qquad (3.30)$$

where $T_\perp^{-1} = 2c_0 + (c_1 + c_{-1})$, $T_\parallel^{-1} = 2(c_1 + c_{-1})$, $c_\infty = 2T_\parallel(c_1 - c_{-1})$. Equation (3.30) describes spin relaxation in an axially-symmetric magnetic field. The parameter $T_\parallel$ ($T_\perp$) is interpreted as the longditudinal (transversal) relaxation time. If $t \to +\infty$ then $\mathfrak{a}_t \to \begin{bmatrix} 0 \\ 0 \\ c_\infty \end{bmatrix}$, and S_t tends to the equilibrium state

$$S_\infty = \frac{1}{2}(I + c_\infty\sigma_3).$$

The generator (3.29) is completely dissipative and this imposes nontrivial restrictions on the physical parameters of the evolution. In fact $2T_\perp^{-1} - T_\parallel^{-1} = 2c_0 \geq 0$, from whence $2T_\parallel \geq T_\perp$ [84].

In general the following result holds for generators of norm continuous covariant semigroups [121].

Theorem 3.3.5. *Let $g \to V_g$ be a continuous representation of a compact group G. If $\mathcal{L}$ is the generator of a norm continuous covariant dynamical semigroup, then it has a standard representation (3.22) in which Φ is a covariant completely positive map, and $[K, V_g] = 0, g \in G$.*

This result holds for dynamical semigroups on an arbitrary von Neumann algebra $\mathfrak{A}$ and for $\mathfrak{A} = \mathfrak{B}(\mathcal{H})$ it can be described in more detail as follows: there exists a standard representation (3.21) in which $[H, V_g] = 0, g \in G$, and L_j are components of a *tensor operator*, namely, there exists a unitary representation $g \to D_g$ in a Hilbert space $\mathcal{H}_0$ such that

$$V_g^* L_j V_g = \sum_k \langle e_j | D_g e_k \rangle L_k, \tag{3.31}$$

where $\{e_j\}$ is an orthonormal base in $\mathcal{H}_0$. In the case of $G = SO(2)$ one can recover the generator (3.29) of the Bloch equation.

Such decomposition into covariant parts may not hold for unbounded generators and non-compact symmetry groups.

Example 3.3.4. Let $\mathcal{H} = L^2(\mathbb{R})$ be the space of of irreducible representation of the CCR (1.24). Consider the Gaussian dynamical semigroup

$$\Phi_t(X) = \frac{1}{\sqrt{2\pi t}} \int\limits_{-\infty}^{\infty} e^{-\frac{x^2}{2t}} e^{iPx} X e^{iPx} dx, \tag{3.32}$$

which is covariant with respect to the representation $y \to e^{iyQ}$ of the additive group $G = \mathbb{R}$. Its generator

$$\mathcal{L}(X) = PXP - P^2 \circ X,$$

defined on an appropriate domain, does not admit a decomposition into the covariant components.

As explained above in Sect. 3.3.3, the case of unbounded generators is approached more conveniently via Markov master equations, determined by form-generators. Assume that $g \to V_g$ is a representation of the symmetry group G, and let $\mathcal{D}$ be a dense domain in $\mathcal{H}$, invariant under V_g. The form-generator $\mathcal{L}(\psi; X; \phi); \phi, \psi \in \mathcal{D}$, is covariant if $\mathcal{L}(\psi; V_g^* X V_g; \phi) = \mathcal{L}(V_g\psi; X; V_g\phi)$ for all admissible values of the arguments. In this case the minimal solution of the Markov master equation is a covariant dynamical semigroup. Let us denote by

$$\mathfrak{A}_V = \{X : X \in \mathfrak{B}(\mathcal{H}), V_g^* X V_g = X; g \in G\}$$

the *fixed-point algebra* of the representation $g \to V_g$.

Theorem 3.3.6. *The minimal solution Φ_t^∞ is unital if and only if $X_t \equiv I$ is the unique solution of the Cauchy problem*

$$\frac{d}{dt}\langle \psi | X_t \psi \rangle = \mathcal{L}(\psi; X_t; \phi); \quad t > 0; \quad X_0 = I \tag{3.33}$$

in the class of uniformly bounded weakly continuous functions with values in $\mathfrak{A}_V$.

In particular, if $\mathcal{L}(\psi; X; \phi) \equiv 0$ for $X \in \mathfrak{A}_V$, then the solution is unique. This is the case when V_g is irreducible, since then $\mathfrak{A}_V = \mathbb{C}I$. Generators of covariant dynamical semigroups acting identically on the fixed-point algebra are described in a series of works of which there is a review in [68]. However this case is too restrictive, and the theorem may be especially helpful in the case where $\mathfrak{A}_V$ is Abelian algebra, since then its condition means essentially nonexplosion of the classical Markov process in $\mathfrak{A}_V$, generated by the backward Kolmogorov equation (3.33).

Example 3.3.5. Consider the form-generator

$$\mathcal{L}(\psi; X; \phi) \equiv \langle P\psi|XP\phi\rangle - \frac{1}{2}\langle P^2\psi|X\phi\rangle - \frac{1}{2}\langle \psi|XP^2\phi\rangle,$$

defined on $\mathcal{D} = C_0(\mathbb{R}^2)$, and corresponding to the formal expression (3.3.4). The fixed-point algebra $\mathfrak{A}_V$ of the representation $y \to e^{iyQ}$ in $L^2(\mathbb{R})$ is the algebra of bounded measurable functions of Q. The Cauchy problem (3.33) reduces to

$$\frac{\partial}{\partial t}X_t(Q) = -\frac{1}{2}\frac{\partial^2}{\partial Q^2}X_t(Q); \quad X_0(Q) = I,$$

which is the Kolmogorov equation for the Wiener process, having the unique solution $X_t(Q) \equiv 1$. Thus (3.32) is the unique solution of the Markov master equation.

Since the properties of symmetry are interesting from both physical and mathematical points of views, it is important to study the structure of Markov master equations covariant with respect to a given symmetry group. This problem was solved for a wide class of groups in [124], and for a particularly important case of the Galilei group in [128]. It was shown that the form-generator admits the standard representation, the components of which obey covariance equations more general than (3.31), in that they also contain a cocycle $\alpha(g)$ of the representation $g \to D_g$ in $\mathcal{H}_0$, that is a solution of the equation $D_g\alpha(h) = \alpha(gh) - \alpha(g); g, h \in G$. The possibility of the simple decomposition of a bounded covariant generator described by theorem 3.3.5 is related to the fact that a bounded cocycle is a coboundary. For many locally compact groups the structure of cocycles is known (see e.g. [89], [186]). This allows us to solve the covariance equations, leading to a kind of Levy-Khinchin representation for the covariant form-generators. In the case of an Abelian group the covariant dynamical semigroup is an extension to $\mathfrak{B}(\mathcal{H})$ of a classical Markov process with locally independent increments in the fixed-point algebra of the representation, and nonexplosion of this process is the condition for the uniqueness of solution for the quantum backward Markov master equation.

3.3.5 Ergodic Properties

If Φ is a dynamical map, then the family $\{\Phi^k; k = 0, 1, \ldots\}$ can be considered as a dynamical semigroup in discrete time. The asymptotic properties of such semigroups as $k \to \infty$ are a nontrivial generalization of the ergodic theory for classical Markov chains. The property of complete positivity is used here only in so far as it implies the Kadison-Schwarz inequality (3.2). Mean ergodic theorems hold for positive maps.

In the majority of works it is assumed that Φ has a faithful normal invariant state, i.e. a state with nondegenerate density operator S_∞ such that $\operatorname{Tr} S_\infty \Phi[X] = \operatorname{Tr} S_\infty X$ for all $X \in \mathfrak{B}(\mathcal{H})$. The mean ergodic theorem,

$$ w^* - \lim_{k \to \infty} \frac{1}{k} \sum_{j=0}^{k} \Phi^k[X] = \mathcal{E}_\infty[X], \qquad (3.34) $$

holds, where $\mathcal{E}_\infty$ is the conditional expectation onto the subalgebra $\mathfrak{A}_\infty$ of elements invariant under Φ. In different degrees of generality this result was obtained by Sinai, Morozova and Chentsov, Kümmerer and other authors (see the review [178]). The relation (3.34) can be regarded as a generalization of the law of large numbers. Much attention has been given to extending theorems of the type (3.34) to unbounded operators and to the study of noncommutative analogues of almost sure convergence in von Neumann algebras. A detailed review of these results is given in [140].

The map Φ is *irreducible* if there does not exist a projector $P \neq 0, I$, such that $\Phi[P] = P$. The last relation is equivalent to the requirement that the subalgebra of operators of the form PXP; $P \in \mathfrak{B}(\mathcal{H})$, is invariant under Φ. Irreducibility is equivalent to the uniqueness of the invariant state S_∞ and the fact that the subalgebra $\mathfrak{A}_\infty$ is one-dimensional. Then $\mathfrak{A}_\infty[X] = (\operatorname{Tr} S_\infty X) \cdot I$ in (3.34). If Φ is written in the form (3.4) then for the irreducibility it is necessary and sufficient that $\{V_n; n = 1, 2, \dots\}'' = \mathfrak{B}(\mathcal{H})$, where $\mathfrak{A}'' = (\mathfrak{A}')'$ is the von Neumann algebra generated by $\mathfrak{A}'$. For an irreducible map there is an analogue of the Perron-Frobenius theorem, corresponding to the decomposition of a closed class of states of a Markov chain into subclasses (see the review [178]). The case $\dim \mathcal{H} < \infty$ is also studied in detail in the book by Sarymsakov [201].

Now let $\{\Psi_t; t \in \mathbb{R}_+\}$ be a quantum dynamical semigroup, having a faithful normal invariant state S_∞. The mean ergodic theorem also holds for this case (see the reviews [178], [87]) . Irreducible semigroups have been studied by Davies, Evans, Spohn and Frigerio, see [56], [209]. A necessary and sufficient condition for irreducibility of the dynamical semigroup with the generator (3.21) is that

$$ \{H, L_j L_j^*; j = 1, 2, \dots\}'' = \mathfrak{B}(\mathcal{H}). $$

For irreducible quantum dynamical semigroups with continuous time there is an essentially stronger form of the ergodic theorem (see [87])

$$w^* - \lim_{t \to +\infty} \Phi_t[X] = (\operatorname{Tr} S_\infty X) \cdot I.$$

This result does generalize to dynamical semigroups in an arbitrary von Neumann algebra. The generalizations to this case of the other asymptotic properties, the spectral theory and the Perron-Frobenius theorem is considered in detail in the review [87].

Important physical examples of ergodic dynamical semigroups arise in the class of quasifree semigroups, for which ergodicity can be established directly (see the review [178]).

3.3.6 Dilations of Dynamical Semigroups

From the point of view of statistical mechanics it is natural to ask, to what extent the concept of dynamical semigroup is compatible with the more fundamental law of irreversible evolution for isolated system. In physical applications the master equation (3.34) is obtained by considering the coupling of the quantum system to the environment in the Markov approximations (weak or singular coupling limits). A rigorous formulation for such an approximation requires rather involved calculations even for simple models. In a number of reviews [209], [10], [84], [178], [1], [3] this problem of the quantum statistical mechanics has been examined from different points of view, and the reader is referred to one these reviews.

Also of interest is the inverse problem of the dilation of a dynamical semigroup to a group of automorphisms, that is the representation of Markov dynamics by reversible dynamics of a system interacting with its environment. The key condition for the existence of such a dilation is the property of complete positivity [57], [69].

Theorem 3.3.7. *Let $\{\Psi_t;\ t \in \mathbb{R}_+\}$ be a norm continuous dynamical semigroup in the state space $\mathfrak{S}(\mathcal{H})$. There exist a Hilbert space $\mathcal{H}_0$, a state S_0 in $\mathcal{H}_0$ and a strongly continuous group of unitary operators $\{U_t;\ t \in \mathbb{R}_+\}$ in $\mathcal{H} \otimes \mathcal{H}_0$, such that*

$$\Psi_t[S] = \operatorname{Tr}_{\mathcal{H}_0} U_t(S \otimes S_0)U_t^*;\ S \in \mathfrak{S}\mathcal{H},$$

for all $t \in \mathbb{R}_+$.

In Sect. 5.2.2 of Chap. 5 we shall give an explicit construction of the dilation which admits a clear dynamical interpretation. We note that the same dynamical semigroup may have many inequivalent dilations (even if one requires "minimality" of the dilation). The concept of quantum stochastic process throws additional light on the structure of possible dilations. According to the definition of Accardi, Frigerio and Lewis [1] a *quantum stochastic process* is defined as a triple $(\mathfrak{A}), (j_t), \varphi)$, where $\mathfrak{A}$ is a C^*-algebra, $(j_t; t \in \mathbb{R})$ is a family of *-homomorphisms from some fixed C^*-algebra $\mathfrak{B}$ into $\mathfrak{A}$, and φ is a state on $\mathfrak{A}$. Under certain regularity conditions a quantum stochastic process is recovered uniquely, to within equivalence, from the *correlation kernels*

$$w_{t_1,\ldots,t_n}(X_1,\ldots,X_n;\ Y_1,\ldots,Y_n)$$
$$= \varphi(j_{t_1}(X_1)^* \ldots j_{t_n}(X_n)^* j_{t_n}(Y_n) \ldots j_{t_1}(Y_1))$$

(this is a noncommutative analogue of Kolmogorov's extension theorem for a system of finite dimensional distributions). In the classical case, $\mathfrak{A}$ and $\mathfrak{B}$ are commutative algebras of measurable bounded functions on the space of events Ω and on the phase space E of the system, respectively, (j_t) is defined by a family of random variables on Ω with values in E, and φ is the expectation functional, corresponding to a probability measure on Ω.

With a quantum stochastic process one associates the families of the "past", "present" and "future" subalgebras

$$\mathfrak{A}_{t]} = \bigvee_{s \leq t} j_s(\mathfrak{B}), \quad \mathfrak{A}_t = j_t(\mathfrak{B}), \quad \mathfrak{A}_{[t} = \bigvee_{s \geq t} j_s(\mathfrak{B}).$$

The process is called *Markov process* if there exists a family of expectations $(\mathcal{E}_{t]}, t \in \mathbb{R})$ from $\mathfrak{A}$ onto $\mathfrak{A}_{t]}$, compatible with φ, such that

$$\mathcal{E}_{t]}(\mathfrak{A}_{[t}) \subseteq \mathfrak{A}_t,$$

a *covariant Markov process* if there exists a group of *-automorphisms $(\alpha_t, t \in \mathbb{R})$ of the algebra $\mathfrak{A}$ such that $\alpha_t(\mathfrak{A}_{s]}) = \mathfrak{A}_{t+s]};\ t,\ s \in \mathbb{R}$ and $\alpha_t \cdot \mathcal{E}_{s]} \cdot \alpha_{-t} = \mathcal{E}_{t+s]}$, and finally a *stationary Markov process* if in addition φ is invariant with respect to (α_t). For a covariant Markov process, the relation

$$\Phi_t[X] = j_t^{-1}\mathcal{E}_{t]}j_t[X] \tag{3.35}$$

defines a dynamical semigroup in $\mathfrak{B}$, provided certain continuity conditions hold. Conversely, any norm continuous quantum dynamical semigroup $\{\Phi_t\}$ can be dilated to a covariant Markov process, satisfying (3.35). Moreover, if the dynamical semigroup satisfies the condition of detailed balance then it can be extended to a stationary Markov quantum stochastic process. The *detailed balance condition* for the state S and the semigroup $\{\Phi_t\}$ in $\mathfrak{B}(\mathcal{H})$ means that there is another dynamical semigroup $\{\Phi_t^+\}$ in $\mathfrak{B}(\mathcal{H})$ such that

$$\mathrm{Tr}\ S\Phi_t^+[X]Y = \mathrm{Tr}\ SX\Phi_t[Y];\ X, Y \in \mathfrak{B}(\mathcal{H}),$$

with infinitesimal generator $\mathcal{L}^*$, satisfying

$$\mathcal{L}[X] - \mathcal{L}^+[X] = 2i[H, X],$$

where $H \in \mathfrak{B}_h(\mathcal{H})$ (the state S is then necessarily stationary for Φ_t and Φ_t^+).

Frigerio and Maassen [73] found a large class of semigroups not satisfying the conditions of detailed balance, but admitting a dilation with the "quantum Poisson process", Kümmerer [152] undertook a systematic study of stationary Markov dilations. He established a direct connection between ergodic properties of a dynamical map (irreducibility, weak and strong mixing) and of its minimal stationary Markov dilation. Kümmerer and Maassen [153] showed that a quantum dynamical semigroup in a finite dimensional Hilbert space admits a stationary Markov dilation using a classical stochastic process if and only if its generator has the form

$$\mathcal{L}[X] = i[H, X] + \sum_{s}(A_s X A_s - A_s^2 \circ X)$$
$$+ \sum_{r} \lambda_r (U_r^* X U_r - X), \tag{3.36}$$

where the A_s are Hermitian, U_r are unitary operators, and $\lambda_r > 0$. The operator (3.36) is the sum of expressions (3.18), (3.19), corresponding to Gaussian and Poisson semigroups, and the dilation is obtained by using a random walk over the group of automorphisms of the algebra $\mathfrak{M}_n$. Similar generators, but with unbounded A_s, U_r and the second sum replaced with an integral appeared in the study of Galilean covariant dynamical semigroups [128].

The non-uniqueness of the dilation of a dynamical semigroup to a stochastic process is associated with the fact that knowledge of the semigroup $\{\Phi_t\}$ enables us to construct only time-ordered correlation kernels

$$W_{t_1,\ldots,t_n}(X_1,\ldots,X_n; Y_1,\ldots,Y_n) =$$
$$\varphi_0(\Phi_{t_1}[X_1^* \Phi_{t_2-t_1}[\ldots \Phi_{t_n-t_{n-1}}[X_n^* Y_n]\ldots]Y_1]),$$

with $0 < t_1 < \ldots < t_n$ (here $\varphi_0 = \varphi \mid \mathfrak{A}_0$ is the initial state). In classical probability theory the correlation kernels depend symmetrically on the times $t_1,\ldots,t_n$; the Daniel-Kolmogorov construction uniquely associates with the semigroup of transition operators the Markov process, which is its minimal dilation to the group of time shifts in the space of trajectories. A definition of quantum stochastic process based only on time-ordered kernels, was proposed by Lindblad [160], and noncommutative generalizations of the Daniel-Kolmogorov construction were studied by Vincent-Smith [223], Belavkin [24] and Sauvageot [203]. Alicki and Messer established the existence and uniqueness of the solution of a class of *non-linear* kinetic equations, the particular, the quantum Boltzmann equation

$$\frac{dS_t}{dt} = \mathrm{Tr}_{(2)} \, W(S_t \otimes S_t) W^* - (\mathrm{Tr} \, S_t) \cdot S_t, \qquad (3.37)$$

where W is a unitary "pair-collision operator" in $\mathcal{H} \otimes \mathcal{H}$ and $\mathrm{Tr}_{(2)}$ is the partial trace over the second factor in $\mathcal{H} \otimes \mathcal{H}$. Answering a question posed by Streater in [195], Frigerio and Aratari constructed a dilation of the "non-linear quantum dynamical semigroup", defined by equation (3.37), to a unitary evolution in a infinite quantum system with pair interactions (the quantum generalization of the "McKean caricature" of the classical Boltzmann equation). Belavkin [25] gave a construction of a quantum branching process, in which one-particle dynamics is described by a semigroup of nonlinear completely positive maps.

4. Repeated and Continuous Measurement Processes

4.1 Statistics of Repeated Measurements

4.1.1 The Concept of Instrument

Consider successive measurement of the two quantities X and Y, assuming values in the sets X and $\mathcal{Y}$ respectively, of a system in the state S. The joint probability that the outcome x of the first measurement is in the set A, and the outcome y of the second one is in B (where $A \subset X$, $B \subset \mathcal{Y}$) is

$$\mu_S(A; B) = \mu_S(A)\mu_S(B \mid A), \tag{4.1}$$

where $\mu_S(A) = \mu_S(A; Y)$ is the probability that $x \in A$, and $\mu_S(B \mid A)$ is the corresponding conditional probability. Denote by S_A the state of the system after the first measurement (it also depends on S but does not depend on B). Then, according to Equation (2.4) of Chap. 1

$$\mu_S(B \mid A) = \operatorname{Tr} S_A M(B), \tag{4.2}$$

where M is the resolution of the identity corresponding to the observable Y. From (4.1), (4.2) it can be seen that the set function

$$M(A)[S] = \mu_S(A)S_A$$

must be σ-additive in A. This motivates the following definition [59].

Let X be a set with σ-algebra of measurable subsets $\mathcal{B}(X)$. An *instrument* (in the state space) with values in X is a map M defined on $\mathcal{B}(X)$ and satisfying the conditions:

1. $M(\mathcal{B})$ is an operation for arbitrary $B \in \mathcal{B}(X)$;
2. $M(X)$ is a dynamical mapping, i.e. $\operatorname{Tr} M(X)[T] = \operatorname{Tr} T$ for all $T \in \mathfrak{T}(\mathcal{H})$;
3. if $\{B_j\} \subset \mathcal{B}(X)$ is a finite or countable partition of the set $B \in \mathcal{B}(X)$ into pairwise disjoint subsets then

$$M(B)[T] = \sum_j M(B_j)[T], \quad T \in \mathfrak{T}(\mathcal{H}),$$

where the series converges in the norm of $\mathfrak{T}(\mathcal{H})$.

It is postulated that if S is the density operator describing the state of the system before the measurement, then the probability that the measurement outcome lies in the set $B \in \mathcal{B}(X)$ is

$$\mu_S(B) = \operatorname{Tr} \mathcal{M}(B)[S], \qquad (4.3)$$

while the state of that part of the statistical ensemble in which this event occurred is described by the density operator

$$S_B = \mathcal{M}(B)[S] / \operatorname{Tr} \mathcal{M}(B)[S] \qquad (4.4)$$

(provided $\mu_S(B) > 0$). In particular, the change of the state of the whole statistical ensemble is given by the dynamical mapping $S \to \mathcal{M}(X)[S]$.

Passing to dual maps $\mathcal{N}(B) = \mathcal{M}(B)^*$, we obtain a formulation in the algebra of observables: the instrument is a set function $\mathcal{N}$ on $\mathcal{B}(X)$ such that

1. $\mathcal{N}(B)$ is a positive normal map from $\mathfrak{B}(\mathcal{H})$ to itself for arbitrary $B \in \mathcal{B}(X)$;
2. $\mathcal{N}(X)[I] = I$;
3. if $\{B_j\} \subset \mathcal{B}(X)$ is a partition of the set B then $\mathcal{N}(B)[X] = \sum_j \mathcal{N}(B_j)[X]$; $X \in \mathfrak{B}(\mathcal{H})$, where the series converges $*$-weakly.

To each instrument there corresponds a generalized observable

$$M(B) = \mathcal{N}(B)[I]; \quad B \in \mathcal{B}(X), \qquad (4.5)$$

such that

$$\mu_S(B) = \operatorname{Tr} S M(B).$$

Ozawa [181] showed that for an arbitrary instrument $\mathcal{M}$ and an arbitrary state S there exists a family of *posterior states* $\{S_x; x \in X\}$, i.e. of density operators S_x, such that

1. the function $x \to \operatorname{Tr} S_x Y$ is μ_S-measurable for arbitrary $Y \in \mathfrak{B}(\mathcal{H})$;
2. $\operatorname{Tr} \mathcal{M}(B)[S]Y = \int_B (\operatorname{Tr} S_x Y) \mu_S(dx)$

The density operator S_x describes the state of the part of the statistical ensemble in which the outcome of the measurement is x, and the quantity $\operatorname{Tr} S_x Y = \mathbf{E}_S(Y|x)$ is the *posterior mean* of the observable Y under the condition that the outcome of the preceding measurement is x.

The instrument $\mathcal{M}$ (or $\mathcal{N}$) is said to be *completely positive* if the maps $\mathcal{N}(\mathcal{B}); B \in \mathcal{B}(X)$ are completely positive.

Example 4.1.1. Let $A = \sum_i x_i E_i$ be a real observable with purely discrete spectrum. The relation

$$\mathcal{M}(B)[S] = \sum_{i:x_i \in B} E_i S E_i \qquad (4.6)$$

defines a completely positive instrument with values in $\mathbb{R}$, corresponding to von Neumann's *projection postulate*, which describes an exact measurement of the observable A. The probability distribution in the state S is

$$\mu_S(B) = \sum_{i:x_i \in B} \operatorname{Tr} S E_i,$$

and the posterior states are given by

$$S_i = E_i S E_i / \operatorname{Tr} S E_i. \qquad (4.7)$$

Example 4.1.2. Let A be a real observable and let $p(x)$ be a probability density on $\mathbb{R}$ such that

$$\int_{-\infty}^{\infty} x p(x) dx = 0, \quad \int_{-\infty}^{\infty} x^2 p(x) dx = \sigma^2 < \infty. \qquad (4.8)$$

The completely positive instrument

$$\mathcal{M}(B)[S] = \int\int_B \sqrt{p(xI - A)} S \sqrt{p(xI - A)} dx \qquad (4.9)$$

describes an imprecise measurement of the observable A with a random error, distributed with the density $p(x)$. In fact, the probability distribution of the outcomes is

$$\mu_S(B) = \int_B \operatorname{Tr} S p(xI - A) dx =$$
$$= \int\int_B p(x - y) \mu_S^A(dy) dx,$$

where μ_S^A is the probability distribution of the observable A in the state S. The posterior states are

$$S_x = \frac{\sqrt{p(xI - A)} S \sqrt{p(xI - A)}}{\operatorname{Tr} S p(xI - A)}.$$

The smaller σ^2 is, that is the closer $p(x)$ to a δ-function, the more accurate is the measurement of A. If A has a purely discrete spectrum, then the case $\sigma^2 = 0$ corresponds to example 4.1.1.

For an observable with continuous spectrum, there are fundamental difficulties preventing a direct generalization of the projection postulate (see Sect. 4.1.4 below).

4.1.2 Representation of a Completely Positive Instrument

Many real processes are described by the following scheme of indirect measurement: the system under investigation interacts with a "probe system", after which a direct measurement of some quantum observable is made on the probe system. Let $\mathcal{H}_0$ be the Hilbert space of the probe system, let S_0 be the density operator describing its initial state, let U be a unitary operator in $\mathcal{H} \otimes \mathcal{H}_0$, describing the interaction and let E_0 be the orthogonal resolution of the identity in $\mathcal{H}_0$ corresponding to the observable. The probability distribution of such a measurement is given by

$$\mu_S(B) = \operatorname{Tr} U(S \otimes S_0)U^*(I \otimes E_0(B)); \quad B \in \mathcal{B}(\mathcal{X}),$$

where S is the initial density operator of the system. It can be expressed in the form (4.3), where

$$\mathcal{M}(B)[S] = \operatorname{Tr}_{\mathcal{H}_0} U(S \otimes S_0)U^*(I \otimes E_0(B)) \qquad (4.10)$$

is completely positive instrument in the space of states of the system $\mathcal{H}$. The converse is also true.

Theorem 4.1.1. *[180] Let $\mathcal{M}$ be a completely positive instrument with values in $\mathcal{X}$. Then there exist a Hilbert space $\mathcal{H}_0$, a density operator S_0 in $\mathcal{H}_0$, a unitary operator U in $\mathcal{H} \otimes \mathcal{H}_0$ and an orthogonal resolution of the identity $\mathcal{E}_0$ in $\mathcal{H}_0$, such that (4.10) holds for an arbitrary density operator S in $\mathcal{H}$.*

This theorem is a combination of Naimark's and Stinespring's theorems: if $\mathcal{N}$ is a completely positive instrument (in an algebra of observables) then there exist a Hilbert space $\mathcal{K}$, an isometric operator V from $\mathcal{H}$ in $\mathcal{K}$, an orthogonal resolution of the identity E in $\mathcal{K}$ and a normal *-homomorphism π from $\mathcal{B}(\mathcal{H})$ into $\mathcal{B}(\mathcal{K})$, such that $[E(B), \pi[X]] = 0$ for all $B \in \mathcal{B}(\mathcal{X})$, $X \in \mathcal{B}(\mathcal{H})$ and

$$\mathcal{N}(B)[X] = V^* E(B)\pi[X]V. \qquad (4.11)$$

The space $\mathcal{K}$ is transformed into $\mathcal{H} \otimes \mathcal{H}_0$ by using the argument of the proof in Sect. 3.1.1, Chap. 3.

Example 4.1.3. Consider the completely positive instrument (4.9) describing approximate measurement of quantum observable A. Let $\mathcal{H}_0 = \mathrm{L}^2(\mathbb{R})$ be the space of the Schrödinger representation Q, P of CCR, which will describe the probe system. As the initial state of the probe system we take $S_0 = |\psi_0\rangle\langle\psi_0|$, where $\psi_0(x) = \sqrt{p(x)}$. Let $U_t = \exp(-itH_{int})$ be the unitary evolution in $\mathcal{H} \otimes \mathcal{H}_0$, where

$$H_{int} = \lambda(A \otimes P),\tag{4.12}$$

and λ is chosen large enough in order to neglect the free dynamics of the system and of the probe. Finally, let $E_Q = |x\rangle\langle x|dx$ be the spectral measure of the position observable Q of the probe system. Then for $t = \frac{1}{\lambda}$ the relation (4.10) holds for the instrument (4.9). The proof is based on the formula

$$\langle x|e^{-i(A\otimes P)}\psi_0\rangle = \psi_0(xI - A),$$

which follows from the fact that $P = -i\frac{d}{dx}$.

This scheme is a generalization of the von Neumann measurement process for an observable A with purely point spectrum (see [175], Ch IV, §3). It reduces approximate measurement of an arbitrary observable A to a position measurement of the probe system.

An important example of a completely positive instrument is

$$\mathcal{N}(B)[X] = \int_B V(x)^* X V(x)\mu(dx),$$

where μ is a σ-finite measure on $\mathcal{X}$ and $V(x)$ is a μ-measurable function on $\mathcal{X}$ with values in $\mathcal{B}(\mathcal{H})$ such that

$$\int_{\mathcal{X}} V(x)^* V(x)\mu(dx) = 1.$$

The corresponding instrument in the space of states has the form

$$\mathcal{M}(B)[S] = \int_B V(x)SV(x)^*\mu(dx).\tag{4.13}$$

Here the probability distribution in the state S is

$$\mu_S(B) = \int_B \operatorname{Tr} SV(x)^* V(x)\mu(dx),\tag{4.14}$$

and the posterior states are

$$S_X = V(x)SV(x)^*/\operatorname{Tr} SV(x)^* V(x).$$

By using the representation (4.11), it is possible to show that this example is generic in the sense that an arbitrary completely positive instrument can be represented as a sum of terms of the form (4.13), where, however, the $V(x)$ are in general unbounded operators (see Sect. 4.1.6).

4.1.3 Three Levels of Description of Quantum Measurements

The theorem of Sect. 4.1.3 is important because it demonstrates the compatibility of the concept of a (completely positive) instrument with the standard formalism of quantum mechanics. The description of the measurement in the generalized statistical model of quantum mechanics can be realized with a different degree of detail. There are three fundamental levels of description, to each of which corresponds a definite mathematical object in the Hilbert space of the system.

1. Only the statistics of the outcomes of the measurement is given. As shown in Sect. 2.1.2 of Chap. 2 this is equivalent to giving a generalized observable, that is, a resolution of the identity in $\mathcal{H}$.
2. In addition to the statistics, the law of the transformation of states depending on the outcome of the measurement is also given. At this level an adequate description of the measurement is given by the concept of instrument. According to (4.5), to each instrument there corresponds a generalized observable, however, the correspondence is not one-to-one, since the instrument gives a more detailed description of the measurement than does the observable.
3. A dynamical description of the interaction of the system with the apparatus is given. This level is yet more detailed; according to (4.10), to each instrument there may correspond many different measurement processes depending on where in the measuring apparatus the distinction is made between the "probe system" and the "detector" realizing the direct measurement.

From the point of view of physical applications, the question of the realizability of the different theoretical quantum measurement schemes is of great interest. An opinion has been expressed (see for example, the article "The problem of measurement" in the collection of Wigner's works [230]), that although quantum mechanics adequately reflects some features of the microscopic world, not all the aspects of the mathematical model necessarily have their prototypes in reality. For example, restrictions of the type of superselection rules are well known (see, for example, [162]); they postulate that only those observables are measurable which are compatible with some selected quantities of the type of electric charge. The following questions arise in connection with indirect measurements:

1. Does a given unitary operator U correspond to a real quantum interaction?
2. Does a given observable correspond to a really measurable physical quantity?

A detailed discussion of these questions goes beyond the framework of the present review but some comments are necessary. In the papers of Waniewski [225] and Mielnik [171] it is shown that every unitary operator may be obtained from Schrödinger evolutions with a time-dependent potential. Thus the first question reduces to the realizability of such potentials in quantum mechanics. On the other hand it will be shown in Sect. 4.5 that the second question reduces to the first one for interactions of a special form. In practice, of course, these questions may be anything but simple and require a separate study for every concrete measurement problem (see, for example, an instructive discussion of the "Dolinar receiver" in [94] and [107]). A completely new approach to that problem is suggested by quantum computing: it is shown that arbitrary unitary operator in a finite-dimensional Hilbert space can be approximated by a finite array of elementary "gates" involving at most two qubits (see e. g. [212], [145]).

4.1.4 Repeatability

Let S be a density operator, and let $\mathcal{M}_1, \dots, \mathcal{M}_n$ be a sequence of instruments with values in the measurable spaces $\mathcal{X}_1, \dots, \mathcal{X}_n$. From the postulates (4.3), (4.4) it follows that the value

$$\mu_S^{\mathcal{M}_1, \dots \mathcal{M}_n} (B_1 \times \dots \times B_n) =$$
$$= \operatorname{Tr} \mathcal{M}_n(B_n)[\dots \mathcal{M}_1(B_1)[S]\dots], \tag{4.15}$$

where $B_j \in \mathcal{B}(\mathcal{X}_j)$ is the probability of the outcomes $x_j \in B_j$; $j = 1, \dots, n$ in the sequence of measurements described by the instruments $\mathcal{M}_1, \dots, \mathcal{M}_n$, over a system initially in the state S. If $\mathcal{X}_1, \dots, \mathcal{X}_n$ are standard measurable spaces, then the set function (4.13), defined on the parallelepipeds $B_1 \times \dots \times B_n$, extends uniquely to a probability measure on the σ-algebra $\mathcal{B}(\mathcal{X}_1) \times \dots \times \mathcal{B}(\mathcal{X}_n)$, see [56] §42. The relation (4.15) can also be written as

$$\mu_S^{\mathcal{M}_1, \dots \mathcal{M}_n} (B_1 \times \dots \times B_n) =$$
$$= \operatorname{Tr} S \mathcal{N}_1(B_1)[\dots \mathcal{N}_n(B_n)[I]\dots], \tag{4.16}$$

In the case of instruments corresponding to the projection postulate (4.6), the relation (4.15) transforms into the Wigner formula (see the article "The problem of measurement" in [230]).

Consider the repeated measurement described by the instrument $\mathcal{M}$. The instrument $\mathcal{M}$ is called *repeatable* if

$$\mathcal{M}(B_1)[\mathcal{M}(B_2)[S]] = \mathcal{M}(B_1 \cap B_2)[S]; \ B_1, \ B_2 \in \mathcal{B}(\mathcal{X}) \tag{4.17}$$

for an arbitrary density operator S. This property is a mathematical expression of the *repeatability hypothesis*, saying that "if a physical quantity is measured twice on the system $\mathcal{S}$, the measurements following directly one after another, then the same value is obtained in both cases" (see [175] Chap. 4, Sect. 3).

Consider an instrument with a countable set of outcomes $X = \{x_1, x_2, \dots\}$ and put $M_i = M(\{x_i\})$.

Proposition 4.1.1. *[180, 56]. Any instrument of the form* (4.6) *has the properties*

1. $M_i[M_j[S]] = \delta_{ij} M[S]$ *(repeatability)*;
2. *If* $\operatorname{Tr} M_i[S] = 1$, *then* $M_i[S] = S$ *(minimality of perturbation)*.
3. *If* $X \geq 0$ *and* $M_i^*[X] = 0$ *for* $i = 1, 2, \dots$, *then* $X = 0$ *(non-degeneracy)*.

Conversely, every instrument with these properties has the form (4.6).

Thus the projection postulate (4.6) can be regarded as the consequence of a number of physically significant properties of the corresponding instrument, including repeatability. As we have already noted, in the case of continuous spectrum difficulties arise, which are most clearly expressed by the following theorem.

Theorem 4.1.2. *[180] Let* X *be a standard measurable space. Any instrument with values in* X *having the property of repeatability* (4.17), *is necessarily discrete, that is, there exists a countable subset* $X_0 \subset X$, *such that* $M(X \setminus X_0)[S] = 0$ *for all* S.

Proof. Consider the faithful state given by a nondegenerate density operator S. According to Sect. 4.1, there exists a family of posterior states $\{S_x\}$. Let $\{B_n\}$ be a countable subalgebra generating $\mathcal{B}(\mathcal{H})$. Introducing the notation $M(B) = M^*[I]$ and using repeatability we have

$$\int_B \operatorname{Tr} S_x M(B_n) \mu_S(dx) = \operatorname{Tr} M(B)[S] M(B_n) =$$

$$= \operatorname{Tr} M(B \cap B_n)[S] = \int_B 1_n(x) \mu_S(dx),$$

from whence $\operatorname{Tr} S_x M(B_n) = 1_{B_n}(x)$ for μ_S-almost all $x \in X$. Therefore there exists a subset $X_0 \subset X$, such that $M(X/X)[S] = 0$ and

$$\operatorname{Tr} S_x M(B) = 1_B(x); \quad x \in X_0, \quad B \in \mathcal{B}(X).$$

But then $\operatorname{Tr} S_x M(\{x\}) = 1$, i.e. $M(\{x\}) \neq 0$ and $\mu_S(\{x\}) \neq 0$ for $x \in X_0$, whence it follows that the X_0 is countable.

In example 4.1.1 the posterior states (4.7) are such that in these states the (discrete) observable A has a definite value x_j with probability 1. The origin of the difficulties with a continuous spectrum lies in the fact that an (algebraic) state, in which a continuous observable A has a definite value, cannot be normal (that is, it cannot be described by a density operator).

4.1.5 Measurements of Continuous Observables

Srinivas [210] and Ozawa [182] discussed the possibility of describing repeatable measurements of continuous observables by using non-normal states and instruments. Let $A = \int_{-\infty}^{\infty} x E_A(dx)$ be a real observable and let η be an arbitrary invariant mean on the space $C(\mathbb{R})$ of bounded continuous functions on $\mathbb{R}$. Following [210] consider the map $\mathcal{E}_\eta^A$ of the algebra $\mathfrak{B}(\mathcal{H})$ into itself defined by

$$\operatorname{Tr} S \mathcal{E}_\eta^A [X] = \eta_y (\operatorname{Tr} e^{iyA} X e^{-iyA}),$$

where the index y means that the averaging is over the variable y, and introduce the set function

$$\mathcal{N}_A(B)[X] = E_A(B) \mathcal{E}_\eta^A [X]; \quad B \in \mathfrak{B}(\mathbb{R}). \tag{4.18}$$

The map $\mathcal{E}_\eta^A$ is the conditional expectation onto the commutative subalgebra $\{E_A(B); \ B \in \mathfrak{B}(\mathbb{R})\} = \mathfrak{B}_A$, generated by the observable A. If A has a purely point spectrum, then $\mathcal{E}_\eta^A$ is a normal expectation, given by (3.5) in Chap. 3, and the relation (4.18) becomes the projection postulate (4.6). In the general case, the mapping $\mathcal{E}_\eta^A$ is not normal, while the set function (4.18) has all the properties of a repeatable instrument except for normality.

The probabilities of the repeated measurement of observables $A_1, \dots, A_n$ are given by a generalization of formula (4.16):

$$\mu_S^{A_1, \dots, A_n}(B_1 \times \dots \times B_n) =$$
$$= \operatorname{Tr} S \mathcal{N}_{A_1}(B_1)[\dots \mathcal{N}_{A_n}(B_n)[I] \dots].$$

If $A_1, \dots. A_n$ are compatible then $\mathcal{E}_\eta^{A_j}[E_{A_k}(B)] = E_{A_k}(B)$, whence

$$\mu^{A_1, \dots, A_n}(B_1 \times \dots \times B_n) = \operatorname{Tr} S E_{A_1}(B_1) \dots E_{A_n}(B_n), \tag{4.19}$$

that is, probability distribution of repeated exact measurements of compatible observables coincides with the distribution of the exact joint measurement of these observables (see Sect. 1.1.5 in Chap. 1) – a result which confirms the correctness of the "generalized projection postulate" (4.18).

On the other hand, if the A_j are incompatible, then, because of the nonnormality of the maps $\mathcal{E}_\eta^{A_j}$, the set function (4.18) may turn out to be only finitely additive on $\mathfrak{B}(\mathbb{R})$. In order to recover σ-additivity it is necessary to consider the distribution on the compactification of the real line $\bar{\mathbb{R}} = \mathbb{R} \cup \{-\infty\} \cup \{\infty\}$. For example, after an exact measurement of the position observable Q, the system is transformed into a non-normal state in which the momentum P takes the values $\pm\infty$ with positive probabilities [210].

Ozawa [182] constructed a process of indirect measurement corresponding to the generalized projection postulate (4.18) and clarified the role of the invariant mean η. The construction is similar to that of the example from Sect. 4.2, related to approximate measurement of A. Consider the observables Q, P in the auxiliary Hilbert space $\mathcal{H}_0 = L^2(\mathbb{R})$, which will describe the "probe system". Using the Hahn-Banach extension theorem one can show that for an arbitrary state η on the algebra $\mathfrak{B}(\mathcal{H}_0)$ there exists a (non-normal) state E_η, such that

$$E_\eta(f(Q)) = f(0), \quad E_\eta(g(P)) = \eta(g),$$

for arbitrary $f, g \in C(\mathbb{R})$. Let $U_t = \exp(-itH_{int})$ be the unitary evolution in $\mathcal{H} \otimes \mathcal{H}_0$ with the interaction Hamiltonian (4.12). In [182] it is shown that for an arbitrary (normal) state S and arbitrary $X \in \mathfrak{B}(\mathcal{H})$

$$\mathrm{Tr}\, S\mathcal{N}_A(B)[X] = (S \otimes E_\eta)(U_{1/\lambda}^*(X \otimes E_Q(B))U_{1/\lambda}); \quad B \in \mathfrak{B}(\mathbb{R}).$$

Thus the probe system, having an exactly defined position in the state E_η, interacts during the time $1/\lambda$ with the observed system according to the Hamiltonian (4.20), after which an exact measurement of position of the probe system is made.

This scheme reduces the measurement of an arbitrary quantum observable to the measurement of the position of the probe system, provided the interaction Hamiltonian (4.20) is realizable.

4.1.6 The Standard Quantum Limit

It is possible to derive a kind of Stinespring-Kraus representation for a completely positive instrument [132].

Theorem 4.1.3. *Let $\mathcal{N}$ be a completely positive instrument with values in $\mathcal{X}$, then there exist a positive measure on $\mathcal{X}$, a dense subspace $\mathcal{D} \subset \mathcal{H}$ and a family $\{V_k(x); k = 1, 2, \ldots; x \in \mathcal{X}\}$ of operators defined on $\mathcal{D}$ such that*

$$\langle \psi | \mathcal{N}(B)[X]\psi \rangle = \int_B \sum_k \langle V_k(x)\psi | X V_k(x)\psi \rangle \mu(dx); \quad \psi \in \mathcal{D}. \qquad (4.20)$$

The operators $V_k(x)$ must satisfy the normalization condition

$$\int_{\mathcal{X}} \|V_k(x)\psi\|^2 \mu(dx) = \|\psi\|^2; \quad \psi \in \mathcal{D}. \qquad (4.21)$$

The representation (4.20) is derived from (4.11), by using von Neumann's spectral theorem for the spectral measure $E(B) = I \otimes E_0(B)$ and decomposing the Hilbert space $\mathcal{H}_0$ into a direct integral with respect to a positive measure μ. Note that the operators $V_k(x)$ need not be bounded or even closable in this representation.

Example 4.1.4. Let $\mathcal{H} = L^2(\mathbb{R})$ be the space of Schrödinger representation of CCR, let $V(x) = |\psi_x\rangle\langle x|$, where $\psi_x; x \in \mathbb{R}$, is a family of unit vectors in $\mathcal{H}$, and let $\mathcal{D} \in \mathcal{H}$ be the subspace of continuous functions. Thus

$$\|V(x)\psi\|^2 = |\langle x|\psi\rangle|^2, \quad \psi \in \mathcal{D},$$

so that (4.21) holds with $k = 1, \mu(dx) = dx$. The relation (4.20) then defines a completely positive instrument, such that

$$\mathcal{M}(dx)[S] = |\psi_x\rangle\langle x|S|x\rangle\langle\psi_x|dx, \tag{4.22}$$

with the probability of outcomes

$$\mu_S(dx) = \langle x|S|x\rangle dx = \text{Tr}\, SE_Q(dx),$$

and the posterior states $\psi_x; x \in \mathbb{R}$. It describes precise measurement of Q, leaving the system in the state ψ_x, if the measurement outcome was x. The operators $V(x)$ are nonclosable, their adjoints having the trivial domain. The following discussion shows that this is not at all a purely theoretical possibility (see [183] for more detail).

There was a long-standing belief among physicists involved in the problem of detection of gravitation waves, that the repeated position measurements of a free mass such as a gravitation wave interferometer are subject to the so called standard quantum limit (SQL). The latter says that "in the repeated measurement of position of free mass m with the time τ between the two measurements, the results of the second measurement cannot be predicted with uncertainty smaller than $\sqrt{\frac{\hbar\tau}{m}}$". The standard "proof" uses Heisenberg uncertainty relation. Let $Q(t) = Q + Pt/m$ be the position of the mass at time t. Then one argues that

$$\mathbf{D}(Q(\tau)) = \mathbf{D}(Q) + \left(\frac{\tau}{m}\right)^2 \mathbf{D}(P) \geq 2\frac{\tau}{m}\sqrt{\mathbf{D}(Q)\mathbf{D}(P)} \geq \frac{\hbar\tau}{m}. \tag{4.23}$$

However, the first equality and hence the SQL is valid only if Q, P are uncorrelated. Yuen suggested that there could be an arbitrarily precise first position measurement, that leaves the free mass in a "contractive state" with large negative correlation between Q, P. The SQL (4.23) is then not valid and could be breached by the second measurement.

The contractive state can be prepared as follows. Let $\psi_{x,0}$ be the minimum uncertainty state (1.30) with zero mean velocity, and $U_t = \exp\left(-it\frac{P^2}{m}\right)$ be the unitary evolution of the free mass. Then the state vector

$$\psi_x = U_{-\tau}\psi_{x,0} \tag{4.24}$$

evolves during the time τ into the state $\psi_{x,0}$ with the position uncertainty σ^2 which can be made arbitrarily small. Consider the instrument (4.20) with the posterior states (4.24); it describes a precise measurement of Q, leaving the system in the contractive state (4.24). If the second measurement is made at the time τ with sufficiently high precision, it will breach the SQL (4.23).

An explicit realization of this instrument via a measurement process with Hamiltonian quadratic in the canonical variables is described by Ozawa [183].

4.2 Continuous Measurement Processes

4.2.1 Nondemolition Measurements

Consider an isolated quantum system whose evolution is described by the group of unitary operators $\{U_t; t \in \mathbb{R}\}$ in the Hilbert space $\mathcal{H}$. Let $\{A_{jt}; j = 1, \ldots, m; t \in T\}$, where $T \subset \mathbb{R}$, be a family of real observables. Let

$$A_j(t) = U_t^* A_{jt} U_t; \; t \in T, \qquad (4.25)$$

and suppose that for arbitrary times $t_1 < \ldots < t_n$ and arbitrary $j_1, \ldots, j_n$ the observables $A_{j_1}(t_1), \ldots, A_{j_n}(t_n)$ are compatible. According to (4.19), the repeated exact measurement of the observables $A_{j_1}(t_1), \ldots, A_{j_n}(t_n)(j_k = 1, \ldots, m)$ has the probability distribution

$$\mu_S(B_1 \times \ldots \times B_n) = \operatorname{Tr} S E_{(t_1)}(B_1) \ldots E_{(t_n)}(B_n), \qquad (4.26)$$

where $B_k \in \mathcal{B}(\mathbb{R}^m)$ and $E_{(t_k)}$ are the spectral measures of the observables $A_{(jt_k)}; j = 1, \ldots, m$.

The family of probability measures (4.26), $n = 1, 2, \ldots; \; t_1, \ldots, t_n \in \mathbb{R}$, is consistent. Using the Kolmogorov extension theorem, it can be shown that there exists a unique orthogonal resolution of the identity E on $\mathcal{B}(\mathbb{R}^T)$, where $\mathcal{B}(\mathbb{R}^T)$ is the σ-algebra of cylinder sets on the space of trajectories $\mathbb{R}^T$, such that

$$E(x(\cdot) : x(t_1) \in B_1, \ldots, x(t_n) \in B_n) = E_{(t_1)}(B_1) \ldots E_{(t_n)}(B_n).$$

It describes the statistics of an exact measurement, continuous (in time), of the compatible family (4.25), in the sense that the probability of the subset B in the space of trajectories is

$$\mu_S(x(\cdot) \in B) = \operatorname{Tr} S E(B).$$

In the physical literature such measurements, called *nondemolition*, have attracted attention in connection with the problem of detection of a weak force (such as gravitation wave) acting on a probe system [39], [191].

Example 4.2.1. A quantum mechanical oscillator of mass m and frequency ω excited by the scalar force $\phi(t)$ is described by the equations

$$\dot{Q}(t) = m^{-1}P(t), \quad \dot{P}(t) = -m\omega^2 Q(t) + \varphi(t)I, \tag{4.27}$$

where $Q(0) = Q$, $P(0) = P$ are canonical observables, that is $[Q, P] = iI$. Set $A_{1t} = Q\cos\omega t - (P/m\omega)\sin\omega t$, so that

$$A_1(t) = Q(t)\cos\omega t - (P(t)/m\omega)\sin\omega t.$$

From equation (4.27)

$$\dot{A}_1(t) = -(\varphi(t)/m\omega)\sin\omega t I, \tag{4.28}$$

hence the observables $A_1(t)$ are compatible for all t and a continuous nondemolition measurement is possible for the family $\{A_1(t)\}$. The force $\varphi(t)$ can be found by observation of arbitrary trajectory from (4.28).

Similarly, for $A_{2t} = P\cos\omega t + m\omega Q\sin\omega t$ we obtain a family of compatible observables

$$A_2(t) = P(t)\cos\omega t + m\omega Q(t)\sin\omega t,$$

for which $\dot{A}_2(t) = \varphi(t)\cos\omega t I$. We note that $A_1(t)$ and $A_2(t)$ are mutually incompatible, because $[A_1(t), A_2(t)] = iI$. In the spirit of Sect. 2.1.3, Chap. 2, let us consider the family of compatible observables

$$\tilde{A}_1(t) = A_1(t) \otimes I_0 - I \otimes Q_0; \quad \tilde{A}_2(t) = A_2(t) \otimes I_0 + I \otimes P_0 \tag{4.29}$$

in the space $\mathcal{H} \otimes \mathcal{H}_0$, see [116]. The force $\varphi(t)$ is then determined from the relation

$$\varphi(t) = \cos\omega t \tilde{A}_2(t) - m\omega \sin\omega t \tilde{A}_1(t).$$

4.2.2 The Quantum Zeno Paradox

Attempts to study continuous measurements of incompatible observables based on the projection postulate lead to a paradoxical result, at the root of which lies the following mathematical fact. Let H be a selfadjoint operator in $\mathfrak{B}(\mathcal{H})$ and let E be a projector, then

$$\lim_{n\to\infty} (E\exp(itH/n)E)^n = E\exp(itEHE). \qquad (4.30)$$

This follows from the fact that $\|\exp(itH/n) - I - itH/n\| = o(1/n)$ and $E^2 = E$. The generalization of this result to the case of unbounded H is not simple; some conditions were obtained in [72]. These results include the case where $\mathcal{H} = L^2(\mathbb{R}^2)$, $H = -\Delta/2m$ is the free particle Hamiltonian in $\mathbb{R}^n$ and $E = 1_{\mathcal{D}}(\cdot)$ is the indicator of the bounded domain $\mathcal{D} \in \mathbb{R}^n$ with a smooth boundary.

Consider a free particle, evolving on the time interval $[0,t]$, and assume that at each instant of time tk/n, $k = 0,1,\dots,n$, an exact measurement is made of the observable E, described by the projection postulate (4.6). If the outcome of the measurement is 1, this means that the particle is located in the domain $\mathcal{D}$. The probability that the outcome 1 is obtained in all $n+1$ measurements is

$$\mu_S(1,\dots,1) = \mathrm{Tr}(E\exp(itH/n)E)^n S(E\exp(-itH/n)E)^n \qquad (4.31)$$

and as $n \to \infty$, by (4.30) this converges to

$$\mathrm{Tr}\, E\exp(itEHE)S_0 \exp(-itEHE)E = \mathrm{Tr}\, S_0 E,$$

where S_0 is the initial state. If initially the particle is in the domain $\mathcal{D}$, then the probability (4.31) is equal to 1 independently of the evolution, i.e. under continuous exact measurement of position, the particle never leaves the domain $\mathcal{D}$ (see also [56], §7.4). The unusual physical consequences of (4.30) were studied in detail by Misra and Sudarshan [172], who proposed the name "quantum Zeno paradox".

The cause of the paradox is in the fact that the measurement described by the projection postulate, transforming the state of the system into a state corresponding to an exact value of the observable, produces a finite change, which prevails negligibly small time evolution during the time t/n, as $n \to \infty$. To avoid this and to obtain a nontrivial limit process of continuous measurement including the evolution, Barchielli, Lanz and Prosperi [15, 16] proposed studying a sequence of inexact measurement whose accuracy decreases proportionally to the number n of the measurements. The description of the limit process in concrete cases was initially associated with the Feynman integral over trajectories (in this connection see also [169]), however the general picture can be based directly on the ideas outlined in Sect. 4.1, in particular, on the concept of instrument. In the papers [114], [115] a parallel was developed between this approach and classical limit theorems; moreover, the limit process of continuous measurement turned out to be a noncommutative analogue of a process with independent increments. As in the classical case, all such processes are described by a formula of the Levy-Khinchin type [123]. Quantum stochastic processes in the sense of Davies [56] correspond from this point of view to a mixture of Poisson processes. In what follows we briefly describe the results of these papers.

4.2.3 The Limit Theorem for Convolutions of Instruments

For simplicity we restrict ourselves to measurements with values in $\mathbb{R}$ (it matters only that the space of values is a locally compact Abelian group). Let $\mathcal{N}_1, \dots, \mathcal{N}_n$ be instruments (in the algebra of observables) with values in $\mathbb{R}$. There exists a unique instrument $\mathcal{N}$ with values in $\mathbb{R}^n$, such that

$$\mathcal{N}(B_1 \times \dots \times B_n)[X] = \mathcal{N}_1(B_1)[\dots \mathcal{N}_n(B_n)[X] \dots]; \; B_j \in \mathcal{B}(\mathbb{R}).$$

The convolution of the instruments $\mathcal{N}_1, \dots, \mathcal{N}_n$ is defined by the relation

$$\mathcal{N}_1 * \dots * \mathcal{N}_n(B)[X] = \mathcal{N}((x_1, \dots, x_n) : x_1 + \dots + x_n \in B); \; B \in \mathcal{B}(\mathbb{R}),$$

and, according to (4.16), describes the statistics of a sum of outcomes of n repeated measurements, corresponding to the instruments $\mathcal{N}_1, \dots, \mathcal{N}_n$. An instrument $\mathcal{N}$ with values in $\mathbb{R}$ is called *infinitely divisible* if for arbitrary $n = 1, 2, \dots$ there is an instrument $\mathcal{N}_{(n)}$, such that $\mathcal{N} = \mathcal{N}_{(n)} * \dots * \mathcal{N}_{(n)} = \mathcal{N}_{(n)}^{*n}$. The problem of continuous measurement appears to be closely related to the limits of n-fold convolutions of the type $\mathcal{N}_{(n)}^{*n}$ as $n \to \infty$, and with the structure of infinitely divisible instruments. A solution of the problem, describing all possible such limits, is based on a generalization of the method of characteristic functions in probability theory.

Denote by $\mathfrak{F}_\sigma$ the Banach algebra of w^*-continuous linear maps from $\mathfrak{B}(\mathcal{H})$ into itself with the product $\Phi \cdot \Psi[X] = \Phi[\Psi[X]]$. The identity in this algebra is the identity map, denoted Id. We introduce the topology τ in the algebra $\mathfrak{F}_\sigma$ by the family of seminorms

$$\|\Phi\|_S = \sup_{\|X\| \le 1} |\operatorname{Tr} S\Phi[X]], \quad S \in \mathfrak{S}(\mathcal{H}).$$

The *characteristic function* of the instrument $\mathcal{N}$ is defined by

$$\Phi(\lambda)[X] = \int_{\mathbb{R}} e^{i\lambda x} \mathcal{N}(d\lambda)[X], \tag{4.32}$$

where the integral converges in the topology τ. A function $\Phi(\lambda)$ with values in $\mathfrak{F}_\sigma$ is the characteristic function of a completely positive instrument if and only if

1. $\Phi(0)[I] = I$;
2. $\Phi(\lambda)$ is τ-continuous at the point $\lambda = 0$;
3. $\Phi(\lambda)$ is *positive definite* in the following sense: for arbitrary finite collections $\{\psi_j\} \subset \mathcal{H}$, $\{\lambda_j\} \subset \mathbb{R}$, $\{X_j\} \subset \mathfrak{B}(\mathcal{H})$

$$\sum_{j,k} \langle \psi_j \Phi | (\lambda_k - \lambda_j)[X_j^* X_k] \psi_k \rangle \geq 0.$$

(this is an analogue of the Bochner-Khinchin theorem, reducing to it in the case $\dim \mathcal{H} = 1$).

The characteristic function of the convolution $\mathcal{N}_1 * \ldots * \mathcal{N}_n$ is the pointwise product of the characteristic functions $\Phi_1(\lambda) \cdot \ldots \cdot \Phi_n(\lambda)$, so the n-fold convolution $\mathcal{N}_{(n)}^{*n}$ has the characteristic function $\Phi_{(n)}(\lambda)^n$, where $\Phi_{(n)}(\lambda)$ is the characteristic function of the instrument $\mathcal{N}_{(n)}$. The probability distribution $\mu_S^{(n)}$ of the sum of n repeated measurements, described by the instrument $\mathcal{N}_{(n)}$, is defined by

$$\int_{\mathbb{R}} e^{i\lambda x} \mu_S^{(n)}(dx) = \operatorname{Tr} S \Phi_{(n)}(\lambda)^n [I].$$

The following proposition is an analogue of the central limit theorem.

Proposition 4.2.1. *Let $\{\mathcal{N}_{(n)}\}$ be a sequence of completely positive instruments and suppose that the limit*

$$\tau - \lim_{n \to \infty} n(\Phi_{(n)}(\lambda) - \operatorname{Id}) = \mathcal{L}(\lambda) \qquad (4.33)$$

exists. Then the convolution $\mathcal{N}_{(n)}^{*n}$ converges weakly to an infinitely-divisible instrument $\mathcal{N}$ with the characteristic function $\exp \mathcal{L}(\lambda)$ in the sense that

$$\tau - \lim_{n \to \infty} \int_{\mathbb{R}} f(x) \mathcal{N}_{(n)}^{*n}(dx) = \int_{\mathbb{R}} f(x) \mathcal{N}(dx)$$

for all continuous bounded functions $f(x)$.[1]

We note that in general the analogue of the classical condition of asymptotic infinitesimality

$$\lim_{n \to \infty} \|\Phi_{(n)}(\lambda) - \operatorname{Id}\| = 0$$

does not imply (4.33). The problem of describing possible limits $\tau - \lim\limits_{n \to \infty} \Phi_{(n)}(\lambda)$ under this condition remains open.

---

[1] We have in mind the exponential in the Banach algebra $\mathfrak{F}_\sigma$.

Example 4.2.2. Let A, H be (bounded) real observables, let $p(x)$ be a probability density on $\mathbb{R}$ satisfying condition (4.9). Consider the completely positive instrument

$$\mathcal{N}_{(n)}(B)[X] = e^{itH/n} \sqrt{n} \int_B \sqrt{p(\sqrt{n}xI - \tfrac{1}{\sqrt{n}}A)} \times$$

$$\times X \sqrt{p(\sqrt{n}xI - \tfrac{1}{\sqrt{n}}A)} dx e^{-itH/n}. \qquad (4.34)$$

The convolution $\mathcal{N}_{(n)}^{*n}$ has the following statistical interpretation. Consider a quantum system whose dynamics on the interval $[0, t]$ is described by the Hamiltonian H. At the times $t_j = jt/n; j = 0, 1, \dots, n - 1$, an approximate measurement is made of the observable A with the variance $n\sigma^2 = n \int x^2 p(x) dx$, and then the mean $\frac{1}{n} \sum_{j=0}^{n-1} \alpha(t_j^{(n)})$ of the measurement outcomes $\alpha(t_j^{(n)})$ is calculated (which has the variance σ^2). When $n \to \infty$, this tends to the limit $\frac{1}{t} \int_0^t \alpha(\tau) d\tau$, which is the mean of the outcomes $\alpha(\tau)$ of a "continuous measurement" of the observable A. A computation shows that for a sufficiently smooth density $p(x)$ the limit (4.33) equals to

$$\mathcal{L}(\lambda)[X] = it[H, X] +$$

$$+ \frac{1}{4}J(AXA - A^2 \circ X) + i\lambda A \circ X - \frac{1}{2}\sigma^2\lambda^2 X, \qquad (4.35)$$

where $J = \int p'(x)^2 p(x)^{-1} dx$ is the Fischer information for the family of densities $\{p(x + \theta)\}$ with the shift parameter $\theta \in \mathbb{R}$, so that $\sigma^2 J \geq 1$.

4.2.4 Convolution Semigroups of Instruments

The following result of the type of Schoenberg's theorem (see [89], [186]) makes a bridge between conditionally positive definite functions and completely dissipative maps (cf. Sect. 3.3.2 of Chap. 3).

Proposition 4.2.2. *Let $\mathcal{L}(\lambda)$ be a function with values in $\mathfrak{F}_\sigma$. The following conditions are equivalent:*

1. *$\exp t\mathcal{L}(\lambda)$ is positive definite for all $t \geq 0$;*
2. *the function $\mathcal{L}(\lambda)$ is Hermitian, i.e. $\mathcal{L}(-\lambda)[X^*] = \mathcal{L}(\lambda)[X]^*$ and conditionally completely positive, that is, for the arbitrary finite collections $\{\Psi_j\} \subset \mathcal{H}, \{\lambda_j\} \subset \mathbb{R}, \{X_j\} \subset \mathcal{B}(\mathcal{H})$ such that $\sum_j X_j \psi_j = 0$,*

$$\sum_{j,k} \langle \psi_j | \mathcal{L}(\lambda_k - \lambda_j)[X_j^* X_k] \psi_k \rangle \geq 0.$$

The function $\mathcal{L}(\lambda)$ is representable as a limit (4.33) if and only if it satisfies one of the conditions of this proposition, is τ-continuous and $\mathcal{L}(0)[I] = 0$. We call such functions *quasicharacteristic*.

The family of instruments $\{N_t;\ t \in \mathbb{R}_+\}$ forms a *convolution semigroup*, if $N_t * N_s = N_{t+s};\ t,\ S \in \mathbb{R}_+$.

Clearly, all the instruments N_t are infinitely divisible. Let $\Phi_t(\lambda)$ be the characteristic function of the instrument N_t . The relation

$$\Phi_t(\lambda) = \exp t\mathcal{L}(\lambda); \quad t \in \mathbb{R}_+, \tag{4.36}$$

establishes a one-to-one correspondence between quasicharacteristic functions $\mathcal{L}(\lambda)$ and convolution semigroups of completely positive instruments satisfying the continuity condition

$$\lim_{t\to 0}\|N_t(U_0) - \mathrm{Id}\| = 0 \tag{4.37}$$

for an arbitrary neighborhood of zero U_0.

The following result can be considered as a generalization of the Levy-Khinchin representation for the logarithm of the characteristic function of an infinitely divisible distribution.

Theorem 4.2.1. *The function $\mathcal{L}(\lambda)$ with values in $\mathfrak{F}_\sigma$ is a quasicharacteristic function if and only if it admits the representation*

$$\mathcal{L}(\lambda) = \mathcal{L}_0 + \mathcal{L}_1(\lambda) + \mathcal{L}_2(\lambda), \tag{4.38}$$

where $\mathcal{L}_0$ is a completely dissipative mapping of the form (3.3.5), Chap. 3,

$$\mathcal{L}_1(\lambda)[X] = \sigma^2\left[(L^*XL - L^*L \circ X) + i\lambda(L^*X + XL) - \frac{1}{2}\lambda^2 X\right], \tag{4.39}$$

where $\sigma^2 \geq 0$, $L \in \mathfrak{B}(\mathcal{H})$, that is a function of type (4.35), and

$$\mathcal{L}_2(\lambda)[X] = \tag{4.40}$$
$$\int_{\mathbb{R}\backslash 0}\left(\mathcal{J}(dx)[X]e^{i\lambda x} - \mathcal{J}(dx)[I] \circ X - i[X, H(dx)] - X\frac{i\lambda x}{1 + x^2}\mu(dx)\right),$$

where $\mathcal{J}(dx)$ is a $\mathfrak{F}_\sigma$-valued set function such that

$$\int_{\mathbb{R}\backslash 0}\frac{x^2}{1 + x^2}\mathcal{J}(dx)$$

is a bounded map and the condition $\sum_j X_j \psi_j = 0$ *implies*

$$\int_{\mathbb{R}\setminus 0} \sum_{j,k} \langle \psi_j | \mathcal{I}(dx)[X_j X_k] \psi_k \rangle < \infty;$$

the Hermitian operator-valued measure $H(dx) = (F(dx) - F(dx)^*)/2i$, *where* $F(dx)$ *is defined by the relation*

$$\langle \psi_0 | F(dx) \psi \rangle = \langle \psi_0 | \mathcal{I}(dx)[|\psi_0\rangle\langle\phi|]\psi \rangle,$$

with a fixed unit vector ψ_0, *and*

$$\mu(dx) = \langle \psi_0 | \mathcal{I}(dx)[|\psi_0\rangle\langle\psi_0|]\psi_0 \rangle.$$

The proof of (4.38) uses a form of the GNS construction which associates with the conditionally completely positive function $\mathcal{L}(\lambda)$ a pair of commuting cocycles of the algebra $\mathfrak{B}(\mathcal{H})$ and of the group $\mathbb{R}$, and also the information on the structure of such cocycles [89], [69], [52] (the proof can be generalized to an arbitrary von Neumann algebra and an arbitrary Abelian locally compact group, see [18]).

The probabilistic meaning of each of the terms in (4.36) is clarified in connection with continuous measurement processes.

4.2.5 Instrumental Processes

Let $\mathcal{Y}$ be the space of all real functions on $\mathbb{R}$ and let $\mathcal{B}_{a,b}$ be the σ-algebra generated by the increments $y(s) - y(t)$; $a \leq t < s \leq b$. An *instrumental process with independent increments (i-process)* is a family $\{N_{a,b}; a, b \in \mathbb{R}, a \leq b\}$, where $N_{a,b}$ is an instrument (in the algebra of observables) with values in $(\mathcal{Y}, \mathcal{B}_{a,b})$ and

$$N_{a,b}(E) \cdot N_{b,c}(F) = N_{a,c}(E \cap F), \qquad (4.41)$$

if $a \leq b \leq c$ and $E \in \mathcal{B}_{a,b}$, $F \in \mathcal{B}_{b,c}$. If all the instruments $N_{a,b}$ are completely positive then the i-process is said to be *completely positive*. For an arbitrary density operator S and time interval $[a, b]$ the i-process defines the probability distribution

$$\mu_s(E) = \operatorname{Tr} S N_{a,b}(E)[I]; \; E \in \mathcal{B}_{a,b}, \qquad (4.42)$$

on the space of trajectories $y(t)$; $t \in [a, b]$. From the physical point of view, the outcome of continuous measurement is the derivative $\dot{y}(t)$, however it turns out that the distribution (4.42) is concentrated on nondifferentiable functions.

This definition, given in [115], is a modification of Davies' definition [56], using only the space of jump functions and of the more general definition of Barchielli, Lanz and Prosperi [15], based on the spaces of generalized functions. In the case $\dim \mathcal{H} = 1$ the i-processes are ordinary real-valued stochastic processes with independent increments. Every i-process is determined by its finite-dimensional distributions, which by (4.41) have the following structure

$$\mathcal{N}_{\tau_0, \tau_p}(y(\cdot) : y(\tau_1) - y(\tau_0) \in B_1, \ldots, y(\tau_p) - y(\tau_{p-1}) \in B_p) =$$
$$= \mathcal{N}_{\tau_0, \tau_p}(y(\cdot) : y(\tau_1) - y(\tau_0) \in B_1) \cdot \ldots$$
$$\ldots \cdot \mathcal{N}_{\tau_{p-1}, \tau_p}(y(\cdot) : y(\tau_p) - y(\tau_{p-1}) \in B_p),$$

where $\tau_0 \leq \tau_1 \leq \ldots \leq \tau_p$; $B_1, \ldots, B_p \in \mathcal{B}(\mathbb{R})$.

An i-process is said to be *homogenous* if for arbitrary $a, b, \tau \in \mathbb{R}$

$$\mathcal{N}_{a+\tau, b+\tau}(T_\tau(E)) = \mathcal{N}_{a,b}(E), \ E \in \mathcal{B}_{a,b},$$

where $(T_\tau y)(t) = y(t + \tau)$. The relation

$$\mathcal{N}_t(B) = \mathcal{N}_{a,a+t}(y(\cdot) : y(a + t) - y(a) \in B); \ B \in \mathcal{B}(\mathbb{R}), \tag{4.43}$$

defines a convolution semigroup of instruments, in terms of which the finite-dimensional distributions are given by

$$\mathcal{N}_{\tau_0, \tau_p}(y(\cdot) : y(\tau_1) - y(\tau_0) \in B_1, \ldots, y(\tau_p) - y(\tau_{p-1}) \in B_p) =$$
$$= \mathcal{N}_{\tau_1 - \tau_0}(B_1) \cdot \ldots \cdot \mathcal{N}_{\tau_p - \tau_{p-1}}(B_p).$$

If the i-process is continuous in the sense that

$$\lim_{t \to 0} \|\mathcal{N}_{a,a+t}(y(\cdot) : | y(a + t) - y(a)) | \geq \varepsilon)\| = 0 \tag{4.44}$$

for arbitrary $\varepsilon > 0$, then the corresponding convolution semigroup is continuous in the sense of (4.37) and consequently, its characteristic functions have the form (4.36), where $\mathcal{L}(\lambda)$ is a quasicharacteristic function. Conversely, let $\mathcal{L}(\lambda)$ be a quasicharacteristic function, and let $\{\mathcal{N}_t\}$ be the corresponding convolution semigroup, then the relation (4.44) defines finite-dimensional distributions which extend to homogenous, continuous, completely positive i-process $\{\mathcal{N}_{a,b}\}$. Furthermore, by modifying the extension technique of the theory of stochastic processes one can prove that the i-process $\{\mathcal{N}_{a,b}\}$ is concentrated on the subspace $\mathcal{D} \subset \mathcal{Y}$, consisting of functions without discontinuities of the second kind[2]. The function $\mathcal{L}(\lambda)$ is called the *generator* of the i-process $\{\mathcal{N}_{a,b}\}$.

---

[2] See I. I. Gikhman, A.V. Skorokhod — Theory of stochastic processes, Nauka, Moscow, (1973), Vol. 2.

The term $\mathcal{L}_0$ in (4.38) describes the evolution (in general, irreversible) of the quantum system under investigation, occurring independently of the measurement process. In the classical case $(\dim \mathcal{H} = 1)$ this term is absent. The second term $\mathcal{L}(\lambda)$ describes a process of continuous approximate measurement of the observable $A = \sigma^2(L + L^*)$. The corresponding i-process is concentrated on continuous trajectories and corresponds to the classical Wiener process. The multidimensional generalization — the sum of terms of form (4.39) with different operators $L_1, \ldots, L_m$ describes a process of continuous approximate measurement of several (in general incompatible) observables [16]. Finally, the term $\mathcal{L}_2(\lambda)$ presents the jump component of the measurement process.

Example 4.2.3. (see [56] Ch.5.) Let $B \to \mathcal{J}(B)$; $B \in \mathcal{B}(\mathbb{R} \backslash 0)$, in (4.40) be a set function satisfying the definition of a completely positive instrument (see Sect. (4.1)) except for normalization 2). Thus $C = \mathcal{J}(\mathbb{R} \backslash 0)[I]$ is a bounded positive operator. Then the relation

$$\mathcal{L}(\lambda)[X] = \int_{\mathbb{R} \backslash 0} e^{i\lambda x} \mathcal{J}(dx)[X] - C \circ X + i[H, X] \qquad (4.45)$$

defines a quasicharacteristic function. The homogeneous i-process $\{\mathcal{N}_{a,b}\}$ with the generator (4.45) has piecewise-constant trajectories; let, for example, $F \subset \mathcal{D}$ be a subset of trajectories, having exactly m jumps on the segment $[a, b]$. Suppose that the j-th jump occurs on the interval $\Delta_j \subset [a, b]$ and has the size $h_j \in B_j$, where $B_j \in \mathcal{B}(\mathbb{R} \backslash 0)$. If the intervals $\Delta_1, \ldots, \Delta_m$ follow one another without intersections, then

$$\mathcal{N}_{a,b}(F) = \int_{\Delta_1} \cdots \int_{\Delta_m} e^{(t_1-a)\mathcal{L}_0} \cdot \mathcal{J}(B_1) \cdot \ldots \cdot \mathcal{J}(B_m) \cdot$$
$$\cdot \; e^{(b-t_m)\mathcal{L}_0} \, dt_1 \ldots dt_m,$$

where $\mathcal{L}_0[X] = -C \circ X + i[H, X]$.

It is simple to construct the analogue of the Poisson process with the generator

$$\mathcal{L}(\lambda)[X] = \mu(e^{i\lambda h} U^* X U - X) + i[H, X],$$

where U is an isometric operator. This is the *counting process* [211], for which jumps of trajectory have fixed size h and occur at random moments of time according to the exponential distribution with the parameter $\mu > 0$. At the moment of a jump, the state is transformed as $S \to USU^*$, and between the jumps it evolves according to the law

$$S \to e^{-\mu t} e^{-iHt} S e^{iHt}.$$

Let us consider briefly the question of convergence of repeated measurements to a continuous measurement process. Let the time axis $\mathbb{R}$ be decomposed into the intervals $[t_i^{(n)}, t_{i+1}^{(n)}]$ of length $1/n$ and let to each moment of time $t_i^{(n)}$ there correspond a measurement described by the completely positive instrument $\mathcal{N}_{(n)}$ with the characteristic function $\Phi_{(n)}(\lambda)$. A series of such repeated measurements defines naturally the i-process $\{\mathcal{N}_{a,b}\}$, whose trajectories are piecewise-constant functions (see [115]). Denote

$$\mu_{S,X}^{(n)}(B) = \operatorname{Tr} S \mathcal{N}_{a,b}^{(n)}(E)[X]; \; E \in \mathcal{B}_{a,b},$$

where $X \geq 0$.

Theorem 4.2.2. *Let the τ-continuous limit*

$$\tau - \lim_{n \to \infty} n(\Phi_{(n)}(\lambda) - \operatorname{Id}) = \mathcal{L}(\lambda),$$

exist, moreover let $\sup_n \sup_{|\lambda| < 1} n \| \Phi_{(n)}(\lambda) - \operatorname{Id} \| < \infty$. *Then the sequence of measures* $\{\mu_{S,X}^{(n)}\}$ *converges in the Skorokhod topology to the measure*

$$\mu_{S,X}(E) = \operatorname{Tr} S \mathcal{N}_{a,b}(E)[X],$$

where $\{\mathcal{N}_{a,b}\}$ *is a homogenous completely positive i-process with the generator* $\mathcal{L}(\lambda)$.

In particular, the sequence of repeated approximate measurements of the observable A from the example of Sect. 4.2.3 converges to the continuous measurement process with the generator (4.35).

In the case $\dim \mathcal{H} = 1$ this corresponds to Skorokhod's theorem concerning the convergence of sums of independent random values to a process with independent increments.

The knowledge of the i-process enables us, in principle, to determine the characteristics of the probability distribution (4.42) in the space of trajectories, in particular, arbitrary moments [16]. Quantum-statistical applications of the theory under investigation can be based upon this. In [116] a comparison was made of estimates of an unknown force, acting on an open quantum system, evaluated from different continuous measurement processes. Statistics of counting processes was studied in [211]; in [107] an application was considered to optimal discrimination between two coherent states by using photon counting and feedback (the Dolinar receiver). For more recent applications to quantum optics see e. g. [19], [20].

5. Processes in Fock Space

5.1 Quantum Stochastic Calculus

5.1.1 Basic Definitions

Let $\mathfrak{h}$ be a Hilbert space. The *symmetric Fock space* over $\mathfrak{h}$ is defined by

$$\Gamma(\mathfrak{h}) = \sum_{n=0}^{\infty} \oplus \Gamma_n(\mathfrak{h}),$$

where $\Gamma_0(\mathfrak{h}) = \mathbb{C}$, and $\Gamma_n(\mathfrak{h}) = \mathfrak{h}_+^n$ is the symmetric n-fold tensor product of the space $\mathfrak{h}$ (see Sect. 1.3.1 in Chap. 1). $\Gamma_n(\mathfrak{h})$ is called the *n-particle subspace*, and $\Gamma_0(\mathfrak{h})$ *vacuum* subspace. In quantum theory $\Gamma(\mathfrak{h})$ describes a system with a variable (unbounded) number of particles (which are Bosons) [35], [36].

In the case we are interested in, when $\mathfrak{h} = L^2(\mathbb{R}_+)$, the Fock space $\Gamma(\mathfrak{h})$ consists of the infinite sequences

$$\psi = [f_0, f_1(t), \dots, f_n(t_1, \dots, t_n), \dots],$$

where $f_0 \in \mathbb{C}$, $f_n(t_1, \dots, t_n)$ is a complex valued symmetric square-integrable function of $t_1, \dots, t_n \in \mathbb{R}_+$, and

$$\langle \psi | \psi \rangle = \sum_{n=0}^{\infty} \frac{1}{n!} \int_0^{\infty} \cdots \int_0^{\infty} | f_n(t_1, \dots, t_n) |^2 \, dt_1 \dots dt_n < \infty.$$

A convenient modification of this representation was proposed by Maassen in [194]. Let $\tau = \{t_1, \dots, t_n\}$ be a chain in $\mathbb{R}_+$, i.e. a finite subset of $\mathbb{R}_+$, with $| \tau | = n$, ordered so that $t_1 < \dots < t_n$. Denoting by $\mathcal{P}$ the set of all chains of length n, we have $\mathcal{P} = \bigcup_{n=0}^{\infty} \mathcal{P}_n$, where $\mathcal{P}_0 = \{\emptyset\}$. Define a σ-finite measure $\mu(d\tau)$ on $\mathcal{P}$ which coincides with the measure $dt_1 \dots dt_n$ on $\mathcal{P}_n$ for $n > 0$, and for which $\mu(\emptyset) = 1$. For $\psi \in \Gamma(\mathfrak{h})$ we set $\psi(\tau) = f_n(t_1, \dots, t_n)$ if

$|\tau| = n > 0$ and $\psi(\emptyset) = f_0$. The symmetric function f_n is determined by its restriction on $\mathcal{P}_n$, moreover

$$\frac{1}{n!} \int_0^\infty \cdots \int_0^\infty |f_n(t_1, \ldots, t_n)|^2 \, dt_1 \ldots dt_n = \int_{\mathcal{P}_n} |\psi(\tau)|^2 \, \mu(d\tau),$$

so that

$$\langle \psi | \psi \rangle = \sum_{n=0}^\infty \int_{\mathcal{P}_n} |\psi(\tau)|^2 \, \mu(d\tau) = \int_{\mathcal{P}} |\psi(\tau)|^2 \, \mu(d\tau). \qquad (5.1)$$

For arbitrary $t \geq 0$ we define the operators $A(t)$, $A^+(t)$, $\Lambda(t)$ by the relations

$$(A(t)\psi)(\tau) = \int_0^t \psi(\tau \cup \{s\}) ds,$$

$$(A^+(t)\psi)(\tau) = \sum_{s \in \tau} 1_{[0,t]}(s)\psi(\tau \setminus \{s\}),$$

$$(\Lambda(t)\psi)(\tau) = \sum_{s \in \tau} 1_{[0,t]}(s)\psi(\tau) \qquad (5.2)$$

The operator $A(t)$ maps $\Gamma_n \equiv \Gamma_n(\mathrm{L}^2(\mathbb{R}_+))$ into Γ_{n-1}, $A^+(t)$ maps Γ_n into Γ_{n+1} and $\Lambda(t)$ maps Γ_n into Γ_n. In quantum physics $A(t)$ is the (Boson) *annihilation operator* (corresponding to the time segment $[0, t]$), $A^+(t)$ is a *creation operator*, and $\Lambda(t)$ is the (Boson) *particle number*. Their common invariant domain is the subspace Γ_∞, consisting of vectors $\psi \in \Gamma(\mathrm{L}^2(\mathbb{R}_+))$ such that

$$\int_{\mathcal{P}} \lambda^{|\tau|} |\psi(\tau)|^2 \, \mu(d\tau) < \infty$$

for all $\lambda > 0$. The operators $A(t)$, $A^+(t)$ extend uniquely to closed mutually adjoint operators (for which we retain the previous notations). The operators $\Lambda(t)$ and

$$Q(t) = A(t) + A^+(t), \quad P(t) = i(A^+(t) - A(t)) \qquad (5.3)$$

are essentially selfadjoint on Γ_∞ (see, for example, [35], [198]).

From the definitions (5.2) we derive the following commutation relations on Γ_∞

$$[A(t), A(s)] = 0, \qquad\qquad [A^+(t), A^+(s)] = 0,$$
$$[A(t), A^+(s)] \qquad\qquad = (t \wedge s)I,$$
$$[\Lambda(t), \Lambda(s)] \qquad\qquad = 0,$$
$$[\Lambda(t), A(s)] = -A(t \wedge s), \quad [\Lambda(t), A^+(s)] = A^+(t \wedge s), \qquad (5.4)$$

where $t \wedge s = \min(t, s)$. From this it follows that

$$[Q(t), Q(s)] = 0, \qquad\qquad [P(t), P(s)] = 0,$$
$$[Q(t), P(s)] \qquad\qquad = 2i(t \wedge s)I,$$
$$[\Lambda(t), Q(s)] = -iP(t \wedge s), \quad [\Lambda(t), P(s)] = iQ(t \wedge s). \qquad (5.5)$$

Let $f \in L^2(\mathbb{R}_+)$. The corresponding *exponential vector* is the vector $\psi_f \in \Gamma(L^2(\mathbb{R}_+))$ such that $\psi_f(\emptyset) = 1$, $\psi_f(\tau) = \prod_{t \in t} f(t)$. The scalar product of two exponential vectors is

$$\langle \psi_f | \psi_g \rangle = \exp \int_0^\infty \overline{f(t)} g(t) dt.$$

From (5.2) it follows that $A(t)\psi_f = \left(\int_0^t f(s)ds \right) \cdot \psi_f$. The vector ψ_0, corresponding to $f \equiv 0$, is called the *vacuum vector*. It satisfies

$$A(t)\psi_0 = 0, \quad \Lambda(t)\psi_0 = 0. \qquad (5.6)$$

We denote the linear span of the family of exponential vectors by Γ_e. It is dense in $\Gamma(\mathfrak{h})$ (see, for example, [89]).

5.1.2 The Stochastic Integral

One of the fundamental properties of the Fock space is the functorial property

$$\Gamma(\mathfrak{h}_1 \otimes \mathfrak{h}_2) = \Gamma(\mathfrak{h}_1) \otimes \Gamma(\mathfrak{h}_2), \qquad (5.7)$$

In particular, for arbitrary $t \in \mathbb{R}_+$

$$\Gamma(L^2(\mathbb{R}_+)) = \Gamma(L^2(0, t) \otimes \Gamma(L^2(t, \infty)). \qquad (5.8)$$

Here, the exponential vectors, in particular the vacuum, are product vectors:

$$\psi_f = \psi_f^{(0,t)} \otimes \psi_f^{(t,\infty)}. \tag{5.9}$$

Because $L^2(\mathbb{R}_+)$ can be regarded as a continuous direct sum (that is, a direct integral of one-dimensional Hilbert spaces), the Fock space $\Gamma(L^2(\mathbb{R}_+))$ is in a sense a continuous tensor product. This structure underlies the connection between the Fock space, infinite divisibility and processes with independent increments (see, for example, [89], [186]).

In what follows $\mathfrak{H} = \mathcal{H} \otimes \Gamma(\mathcal{L}(\mathbb{R}_+))$, where $\mathcal{H}$ is an "initial" space. Elements of $\mathfrak{H}$ can be regarded as functions $\psi(\tau)$, $\tau \in \mathcal{P}$, taking values in $\mathcal{H}$. It will be convenient, in what follows, not to distinguish between operators, acting in $\mathcal{H}$ or in $\Gamma(L^2(\mathbb{R}_+))$, and their ampliations to $\mathfrak{H}$. For example, $A(t)$ denotes both the operator in $\Gamma(\mathcal{L}^2(\mathbb{R}_+))$ and its ampliation to $\mathfrak{H}$. Denote by $\mathcal{H} \otimes \Gamma_e$ the algebraic tensor product of $\mathcal{H}$ and Γ_e. A family of (in general unbounded) operators $\{M(t); t \in \mathbb{R}_+\}$ defined on $\mathcal{H} \otimes \Gamma_e$, is said to be a *process* in $\mathfrak{H}$.

The relation (5.8), (5.9) define a natural filtration in the space $\mathfrak{H}$. The process $\{M(t); t \in \mathbb{R}_+\}$ in $\mathfrak{H}$ is said to be *adapted* (to the given filtration), if for arbitrary $t \in \mathbb{R}_+$

$$M(t) = M_{t]} \otimes I_{[t}, \tag{5.10}$$

where $M_{t]}$ is an operator acting in $\mathcal{H} \otimes \Gamma(\mathcal{L}^2(0,t))$, and $I_{[t}$ is the identity operator in $\Gamma(\mathcal{L}^2(t,\infty))$. In view of (5.8), (5.9), the expectation $\mathcal{E}_{t]}$ onto an algebra of operators of form (5.10), with respect to the vacuum state $|\psi_0\rangle\langle\psi_0|$ is well defined. The adapted process is said to be a *martingale* if $\mathcal{E}_t[M(s)] = M(t)$ whenever $s > t$. The fundamental processes $\{A(t)\}$, $\{A^+(t)\}$, $\{\Lambda(t)\}$ are martingales.

Hudson and Parthasarathy in [138] constructed a stochastic integral for adapted processes with respect to the fundamental martingales $A^+(t)$, $A^+(t)$, and $\Lambda(t)$. We follow a modified version of it (see author's paper in [197]). The process $\{M(t); t \in [0,T]\}$ is said to be *simple* if there is a partition $0 = t_0 < t_1 < \ldots < t_N = T$, such that $M(t) = M(t_{j-1})$ for $t \in [t_{j-1}, t_j]$. For quadruples of simple adapted processes $\{M(t)_\alpha\}$, $\alpha = 0, 1, 2, 3$, the stochastic integral is defined by

$$I(T) = \int\limits_0^T (M_0 d\Lambda + M_1 dA + M_2 dA^+ + M_3 dt)$$

$$= \sum_{j=1}^N \{M_0(t_{j-1})[\Lambda(t_j) - \Lambda(t_{j-1})]$$

$$+ M_1(t_{j-1})[A(t_j) - A(t_{j-1})] + M_2(t_{j-1})[A^+(t_j) - A^+(t_{j-1})]$$

$$+ M_3(t_{j-1})(t_j - t_{j-1})\}. \tag{5.11}$$

From Journé's inequalities (see [170] V.1.4) we have the following estimate

$$\sup_{0<t<T} \|I(t)\psi \otimes \psi_f\|^2 \leq C(\|f\|) \cdot \left\{ \int_0^T |f(t)|^2 \times \right.$$

$$\times \|M_0(t)\psi \otimes \psi_f\|^2 dt + \int_0^T [\|M_1(t)\psi \otimes \psi_f\|^2 +$$

$$\left. + \|M_2(t)\psi \otimes \psi_f\|^2] dt + \left[\int_0^T \|M_3(t)\psi \otimes \Psi_f\| dt \right]^2 \right\}, \qquad (5.12)$$

where $\psi \in \mathcal{H}$, $f \in L^2(\mathbb{R}_+)$. Let us call $\{M_\alpha(t)\}$ an *admissible* quadruple, if for arbitrary $\varepsilon > 0$ there is a quadruple $\{\tilde{M}_\alpha(t)\}$ of simple adapted processes such that

$$\operatorname*{ess\,sup}_{0\leq t\leq T} \|[M_0(t) - \tilde{M}_0(t)]\psi \otimes \psi_f\| < \varepsilon,$$

$$\int_0^T \|[M_{1,2}(t) - \tilde{M}_{1,2}(t)]\psi \otimes \psi_f\|^2 dt < \varepsilon,$$

$$\int_0^T \|[M_3(t) - \tilde{M}_3(t)]\psi \otimes \psi_f\| dt \quad < \varepsilon,$$

and a *strongly admissible* quadruple, if for arbitrary $\varepsilon > 0$ there is a quadruple $\{\tilde{M}_\alpha\}$ of simple adapted processes with values in $\mathfrak{B}(\mathfrak{H})$ such that

$$\operatorname*{ess\,sup}_{0<t<T} \|[M_0(t) - \tilde{M}_0(t)]\| < \varepsilon,$$

$$\int_0^T \|M_{1,2}(t) - \tilde{M}_{1,2}(t)\|^2 dt \quad < \varepsilon,$$

$$\int_0^T \|M_3(t) - \tilde{M}_3(t)\| dt \quad < \varepsilon. \qquad (5.13)$$

From the inequalities (5.12) it follows that for an arbitrary admissible quadruple the stochastic integral

$$I(T) = \int\limits_0^T (M_0 d\Lambda + M_1 dA + M_2 dA^+ + M_3 dt) \tag{5.14}$$

is defined on $\mathcal{H} \otimes \Gamma_e$ as the strong limit of stochastic integrals of form (5.11) of the simple processes $\{M_\alpha\}$ and is an adapted process. If $M_3 \equiv 0$, then $I(t)$ is a martingale; conversely, it can be shown that a sufficiently regular bounded martingale in Fock space is a stochastic integral (Parthasarathy and Sinha). An example of a bounded martingale which cannot be represented as a stochastic integral with respect to the fundamental processes is contained in Journé's paper [142].

From the definitions (5.2) of the fundamental martingales follows the explicit formula (see the paper of Belavkin in [53])

$$(I(t)\psi)(\tau) = \int\limits_0^t \left[M_3(s)\psi + M_1(s)\psi^{(s)} \right](\tau)ds$$
$$+ \sum_{s \in \tau, s < t} \left[M_2(s)\psi + M_0(s)\psi^{(s)} \right](\tau \backslash \{s\}),$$

where $\psi^{(s)}(\tau) = \psi(\tau \cup \{s\})$, which may serve as an alternative definition stochastic integral, meaningful for a broader class of processes (including nonadapted processes).

Stochastic integrals for processes in antisymmetric Fock space were investigated by Barnett, Streater and Wilde, see [21] and in [193], and by Applebaum and Hudson [137].

5.1.3 The Quantum Ito Formula

The relation (5.14) is conventionally written in the differential form

$$dI = M_0 d\Lambda + M_1 dA + M_2 dA^+ + M_3 dt. \tag{5.15}$$

Let $J(t)$ be another stochastic integral such that $dJ = N_0 d\Lambda + N_1 dA + N_2 dA^+ + N_3 dt$.

Theorem 5.1.1 (Hudson, Parthasarathy [138]). *If the quadruples $\{M_\alpha\}$, $\{N_\alpha\}$ are strongly admissible then the product $I(t)J(t)$ is a stochastic integral, moreover,*

$$d(IJ) = I(dJ) + (dI)J + (dI)(dJ),$$

where the product is evaluted according to the following formal rules: the values at time t of an arbitrary adapted process commute with the stochastic *differentials of the fundamental processes $d\Lambda(t), dA(t), dA^+(t), dt$, while in the term $(dI)(dJ)$ the products of the stochastic differentials of the fundamental processes obey the multiplication table*

$$
\begin{array}{c|cccc}
 & dA^+ & d\Lambda & dA & dt \\
\hline
dA & dt & dA & 0 & 0 \\
d\Lambda & dA^+ & d\Lambda & 0 & 0 \\
dA^+ & 0 & 0 & 0 & 0 \\
dt & 0 & 0 & 0 & 0
\end{array}
\tag{5.16}
$$

It was observed in [117] that the algebra of stochastic differentials (5.15) with the multiplication table (5.16) is isomorphic to an algebra of 3×3-matrices. A symmetric representation was proposed by Belavkin in [53]: the correspondence

$$
dI = M_0 d\Lambda + M_1 dA + M_2 dA^+ + M_3 dt
$$
$$
\longleftrightarrow \quad \mathbf{M} = \begin{bmatrix} 0 & M_1 & M_2 \\ 0 & M_0 & M_2 \\ 0 & 0 & 0 \end{bmatrix}
\tag{5.17}
$$

is a *-algebra isomorphism, taking the involution $(dI)^* = M_0^* d\Lambda + M_2^* dA + M_1^* dA^+ + M_3^* dt$ into the involution

$$
\begin{bmatrix} 0 & M_1 & M_3 \\ 0 & M_0 & M_2 \\ 0 & 0 & 0 \end{bmatrix}^\star = \begin{bmatrix} 0 & M_2^* & M_3^* \\ 0 & M_0^* & M_1^* \\ 0 & 0 & \end{bmatrix}.
$$

Example 5.1.1. Consider the stochastic integrals

$$
B(t) = \int_0^t J(s)dA(s), \quad B^+(t) = \int_0^t J(s)dA^+(s),
$$

where the strongly admissible process $J(t) = (-1)^{A(t)}$. From (5.24) below it follows that $J(t)$ satisfies the equation

$$
dJ = -2Jd\Lambda.
$$

Using the table (5.16) we find that

$$
d(BJ + JB) = -2(BJ + JB)d\Lambda.
$$

Since $B(0) = 0$, we derive from the theorem of the next section that $B(t)J(t) + J(t)B(t) \equiv 0$. Again using the table (5.16), we find that $d(BB^+ + B^+B) = dt$, whence

$$B(t)B^+(t) + B^+(t)B(t) = t,$$

and a similar argument shows that the operators $B(t), B^+(t)$ satisfy the canonical anticommutation relation for Fermion creation-annihilation operators. This fact underlies the isomorphism between symmetric (Boson) and antisymmetric (Fermion) Fock spaces, established by Hudson and Parthasarathy [136].

The quantum stochastic integral and the Ito formula have a natural generalization for many degrees of freedom, when the basic processes $A_j(t), A_k^+(t)$, $\Lambda_{jk}(t)$ are multidimensional and operate in the Fock space $\Gamma(L_{\mathcal{K}}^2(\mathbb{R}_+))$, where $\mathcal{K}$ is a Hilbert space whose dimension is equal to the number of degrees of freedom (Hudson, M.P.Evans in [195], Belavkin in [53]), and also in non-Fock spaces, associated with Gaussian states of canonical commutation relations (Hudson, Lindsay in [194])

5.1.4 Quantum Stochastic Differential Equations

Consider the homogenous linear quantum stochastic differential equation

$$dV = \left[L_0 d\Lambda + L_1 dA + L_2 dA^+ + L_3 dt\right] V, \quad t \geq 0, \qquad (5.18)$$

with the initial condition $V(0) = I$, equivalent to the integral equation

$$V(t) = I + \int_0^t [L_0(s)d\Lambda(s) + L_1(s)dA(s) + L_2(s)dA^+(s)$$
$$+ L_3(s)ds]\, V(s).$$

By modifying the argument of Hudson and Parthasarathy [138] based on the method of successive approximations, the following theorem may be proved

Theorem 5.1.2. *If $\{L_\alpha\}$ is a strongly admissible quadruple, then the solution $\{V(t); t \in \mathbb{R}\}$ of the equation (5.18) exists, is unique and is an adapted process, strongly continuous on $\mathcal{H} \otimes \Gamma_e$.*

Denote by S_t the time-shift operator in $\mathcal{H}: S_t\psi(\tau) = \psi(\tau_t)$, where $\tau_t = \{t_1 + t, \ldots, t_n + t\}$, if $\tau = \{t_1, \ldots, t_n\}$. The solution $V(t)$ satisfies the *cocycle* equation

$$V(t + u) = (S_u^* V(t) S_u)(V(u); \quad t, u \in \mathbb{R}_+. \qquad (5.19)$$

Of particular interest is the case where $V(t)$, $t \in \mathbb{R}_+$, are unitary operators. For this it is necessary and sufficient for equation (5.18) to have the form

$$dV = [(W - I)d\Lambda + LdA^+ - L^*WdA \\ - \left(iH + \tfrac{1}{2}L^*L\right)dt]V,$$

where W is unitary, H is Hermitian and L is an arbitrary bounded operator in $\mathfrak{B}(\mathfrak{H})$.

Of particular importance is the problem of the generalization of the existence and uniqueness theorems for equation (5.18) to the case of unbounded coefficients $L_\alpha(t)$. There are some results in this direction in the papers of Hudson and Parthasarathy [138], Journé [142], Frigerio, Fagnola and Chebotarev [49], Fagnola [70]. In [142] the problem of characterizing strongly continuous unitary solutions of (5.19) is partially solved. This question is closely related to the conservativity problem discussed in 3.3.3.

Equations of the type (5.18) are associated with linear stochastic differential equations in Hilbert space (see, in particular, Skorokhod [208], and these two directions are closely related. Solutions of the equation (5.1.4) are non-commutative analogues of multiplicative processes with independent stationary increments in the groups of unitary operators. The general theory of such processes was developed by Accardi, von Waldenfels and Schürmann [4]. The latter showed in [196] and in [204], that every such process satisfying a condition of uniform continuity is a solution of an equation of the type (5.1.4). Using the analogy with quantum processes, the author exhibited the stochastic differential equation satisfied by an arbitrary classical multiplicative process with independent stationary increments in a Lie group (the multiplicative analogue of the Ito representation [197]).

A natural representation of solutions of equation (5.18) is given by time-ordered exponentials, related to the multiplicative stochastic integral in the classical theory of random processes (concerning the latter see Emery [67]). Let $\{\tilde{M}_\alpha(t)\}$ be a quadruple of simple adapted processes on $[0, T]$ with values in $\mathfrak{B}(\mathfrak{H})$. We set

$$V_j = \exp[\tilde{M}_0(t_{j-1})(\Lambda(t_j) - \Lambda(t_{j-1})) \\ + \tilde{M}_1(t_{j-1})(A(t_j) - A(t_{j-1})) + \tilde{M}_2(t_{j-1})(A^+(t_j) - A^+(t_{j-1})) \\ + \tilde{M}_3(t_{j-1})(t_j - t_{j-1})]$$

and introduce the notation

$$\overleftarrow{\exp} \int_0^T (\tilde{M}_0 d\Lambda + \tilde{M}_1 dA + \tilde{M}_2 dA^+ + \tilde{M}_3 dt) = \\ = V_N \cdot \ldots \cdot V_1. \qquad (5.20)$$

The operator (5.20) is defined on $\mathcal{H} \otimes \Gamma_e$. In the author's papers [120], [127] it is proved that if $\{M_\alpha\}$ is a strongly admissible quadruple of adapted processes with values in $\mathcal{B}(\mathcal{H})$ and $\{\tilde{M}_\alpha^N\}$ in is a sequence of quadruples of simple processes approximating $\{M_\alpha\}$ in the sense (5.13), then the strong limit on $\mathcal{H} \otimes \Gamma_e$ of the expression of form (5.20) exists; it is called the *time-ordered exponential*. Moreover, the family of time ordered exponentials

$$\overleftarrow{\exp} \int_0^T (M_0 d\Lambda + M_1 dA + M_2 dA^+ + M_3 dt), \quad t \in \mathbb{R}_+, \qquad (5.21)$$

is an adapted process strongly continuous on $\mathcal{H} \otimes \Gamma_e$ and satisfies the stochastic differential equation (5.18), where $\{L_\alpha\}$ and $\{M_\alpha\}$ are connected by the relations

$$L_0 = a(M_0), \; L_1 = M_1 b(M_0), \; L_2 = b(M_0) M_2,$$
$$L_3 = M_3 + M_1 c(M_0) M_2.$$

Here a, b, c are the entire functions

$$a(z) = e^z - 1, \; b(z) = \frac{e^z - 1}{z}, \quad c(z) = \frac{e^z - 1 - z}{z^2}, \; z \neq 0. \qquad (5.22)$$

By using the isomorphism (5.17) these relations can be combined into a single matrix equation

$$\mathbf{L} = e^{\mathbf{M}} - I.$$

If the coefficients $M_\alpha(t)$ commute for all times, then the time-ordered exponential transforms into the usual exponent

$$\exp \int_0^t (M_0 d\Lambda + M_1 dA + M_2 dA^+ + M_3 dt), \qquad (5.23)$$

thus giving an explicit solution of (5.18).

Example 5.1.2. The solution of the equation

$$dJ_z = z J_z d\Lambda; \quad J_z(0) = I, \qquad (5.24)$$

for $z \neq -1$ is given by

$$J_z(t) = (z+1)^{A(t)}, \quad t \in \mathbb{R}_+.$$

If $z = -1$ then the solution of example (5.24) has the form

$$J_{-1}(t) = \delta_{0,A(t)},$$

where $\delta_{i,j}$ is the Kronecker symbol. Since $J_{-1}(t)$ can be zero, it cannot be written as an exponential.

The time-ordered exponentials

$$V(t) = \overleftarrow{\exp} \int_0^t (L dA^\dagger - L^* dA - iH dt) \tag{5.25}$$

were considered by von Waldenfels and also by Hudson and Parthasarathy in [193]. These exponentials are unitary operators, satisfying

$$dV = [L dA^+ - L^* dA - \left(iH + \frac{1}{2} L^* L\right) dt]V; \quad V(0) = I \tag{5.26}$$

Example 5.1.3. Let $\dim \mathcal{H} = 1$ and $f \in L^2(\mathbb{R}_+)$. The exponent

$$V_f(t) = \exp \int_0^t (f(s) dA^+(s) - \overline{f(s)} dA(s))$$

is the unitary solution of

$$dV_f(t) = [f(t) dA^+(t) - \overline{f(t)} dA(t) - \frac{1}{2} \mid f(t) \mid^2 dt]V_f(t) \tag{5.27}$$

in $\Gamma(L^2(\mathbb{R}))$. From this equation and the quantum Ito formula, it follows that the processes $V_f(t)V_g(t)$ and $V_{f+g}(t) \exp i\mathrm{Im} \int_0^t f(s)\overline{g(s)} ds$, where g is another element of $L^2(\mathbb{R}_+)$, satisfy the same equation; in addition, they coincide when $t = 0$. Hence they are identically equal, that is,

$$V_f(t)V_g(t) = V_{f+g}(t) \exp \left[i\mathrm{Im} \int_0^t f(s)\overline{g(s)} ds \right]. \tag{5.28}$$

Consider $L^2(\mathbb{R}_+)$ as a real linear space Z with the skew-symmetric form

$$\Delta(f,g) = 2\mathrm{Im}\int_0^\infty f(s)\overline{g(s)}ds.$$

It then follows form (5.28) that the operators $W(f) = V_f(\infty)$ form an irreducible representation of the Weyl canonical commutation relations (1.27) of Chap. 1 in the symmetric (Boson) Fock space $\Gamma(L^2(\mathbb{R}_+))$, and (5.4), (5.5) give the infinitesimal form of the canonical commutation relations. Here the exponential vectors play the same role as coherent states in the case of a finite number degrees of freedom, and the duality map (see (5.2.1)) corresponds to the transformation to the Schrödinger representation, diagonalizing the operators $Q(t)$.

Further information on quantum stochastic calculus can be found in the books by Parthasarathy [185] and Meyer [170] and also in the collections [194]–[197] embracing themes such as the connections with non-commutative geometry (Hudson, Applebaum, Robinson), applications to the theory of multiple stochastic integrals (Maassen, Meyer, Parthasarathy, Lindsay), noncommutative random walks in "toy Fock space" and their convergence to the fundamental processes (Parthasarathy, Lindsay, Accardi) and others.

5.2 Dilations in the Fock Space

Due to its continuous tensor product structure, the Fock space is a natural carrier for various "infinitely divisible" objects. At the end of the 60's Araki and Streater studied infinitely divisible group representations and their embeddings into Fock space. Since a group representation is characterized by a positive definite function, this also gives a representation of an infinitely divisible positive definite function as a Fock vacuum expectation (see [89, 186]). It follows also that infinitely-divisible probability distributions may be realized as distributions of quantum observables in a Fock vacuum state. Thus a Fock space embraces all processes with independent increments and also "quantum noise" processes which give a universal model of the environment of an open quantum system. This fact underlies the construction of dilations using Fock space.

5.2.1 Wiener and Poisson Processes in the Fock Space

If $\{X(t);\ t \in \mathbb{R}\}$ is a commuting family of selfadjoint operators in the Hilbert space $\mathfrak{H}$ then it is *diagonalizable*: there exists a measure space $(\Omega, \mathcal{B}(\Omega)\mu)$ and a unitary operator J from $\mathfrak{H}$ onto $L^2(\Omega, \mu)$, such that

$$(JX(t)J^{-1}\psi)(\omega) == X_t(\omega)\psi(\omega),$$

for $\psi \in L^2(\Omega, \mu)$, where $X_t(\omega)$ are real measurable functions. Moreover for arbitrary $\psi \in \mathfrak{H}$ and a bounded Borel function $f(x_1, \dots, x_n)$,

$$\langle \psi | f(X(t_1), \dots, X(t_n)) \psi \rangle = \int f(X_{t_1}(\omega), \dots, X_{t_n}(\omega)) P(d\omega),$$

where $P(d\omega) = | (J\psi)(\omega) |^2 \mu(d\omega)$ is a probability measure on Ω. In this sense, the family $\{X(t)\}$ in the Hilbert space $\mathfrak{H}$ together with a preferred vector ψ is statistically equivalent to the random process $\{X_t(\omega)\}$ in the probability space $(\Omega, \mathcal{B}(\Omega), P)$.

Consider the family of (commuting by (5.15)) selfadjoint operators $\{Q(t)\}$ in $\Gamma(L^2(\mathbb{R}_+))$. Let $\{W_t; t \in \mathbb{R}_+\}$ be a standard Wiener process, and let $L^2(W)$ be the Hilbert space of complex valued square integrable functions of the Wiener process. The Segal *duality map*

$$J\psi = f_0 + \sum_{n=1} \int \cdots_{\mathfrak{P}_n} \int f_n(t_1, \dots, t_n) dW_{t_1}, \dots, dW_{t_n},$$

where the right hand side comprises multiple stochastic integrals in the Ito sense, is an isomorphism from the Fock space $\mathcal{H} = \Gamma(L^2(\mathbb{R}_+))$ onto $L^2(W)$, moreover

$$J\psi_0 = 1; \quad JQ(t)J^{-1} = W_t.$$

Thus the family $\{Q(t)\}$ in $\Gamma(L^2(\mathbb{R}_+))$ with the vacuum vector ψ_0 is stochastically equivalent to the Wiener process W_t. A similar result also holds for the commuting family $\{P(t)\}$. However by (5.5) the operators $Q(t)$ do not commute with the operators $P(s)$ and therefore the family $\{Q(t), P(t)\}$ is not equivalent to the two-dimensional Wiener process. The unitary operator $U\psi(\tau) = i^{|\tau|}\psi(\tau)$ leaves ψ_0 invariant and satisfies

$$P(t) = UQ(t)U^{-1}.$$

The operator in $L^2(W)$ corresponding to U is the Fourier-Wiener transform[1].

Now consider the commuting family of selfadjoint operators $\{A(t)\}$. Let $\{N_t; t \in \mathbb{R}_+\}$ be a Poisson process of intensity λ on the probability space $(\Omega, \mathcal{B}(\Omega), P_\lambda)$ and let $L^2(N) \equiv L^2(\Omega, P_\lambda)$. The map

---

[1] See T.Hida, Brownian motion, Springer, Berlin, (1980).

$$J^{(\lambda)}\psi = f_0 + \sum_{n=1}^{\infty} \int \cdots_{\mathfrak{P}_n} \int f_n(t_1, \ldots, t_n) dX_{t_1} \ldots dX_{t_n},$$

where $X_t = \lambda^{\frac{1}{2}}(N_t - \lambda t)$ is the compensated Poisson process, is an isomorphism from the Fock space $\Gamma(L^2(\mathbb{R}_+))$ onto $L^2(N)$, moreover

$$J^{(\lambda)}\psi_0 = 1, \quad J^{(\lambda)}\Pi^{(\lambda)}(t)J^{(\lambda)-1} = N_t,$$

where

$$\Pi^{(\lambda)}(t) = A(t) + \sqrt{\lambda}Q(t) + \lambda t. \tag{5.29}$$

Thus, the family $\{\Pi^{(\lambda)}(t)\}$ in $\Gamma(L^2(\mathbb{R}_+))$, together with the vacuum vector ψ_0, is stochastically equivalent to the Poisson process [138].

From the viewpoint of classical probability theory, the relation (5.29) is surprising: the Poisson process is represented as the sum of the Wiener process with constant drift and the process $A(t)$ which vanishes almost surely (in the vacuum state). The point is, of course, that the terms do not commute and therefore cannot be considered as classical random processes on a single probability space. A similar connection between the Poisson and the normal distribution is well known in quantum optics [146].

An arbitrary continuous stochastic process with independent increments may be embedded in a suitable Fock space $\Gamma(L^2_{\mathcal{K}}(\mathbb{R}_+))$, where $L^2_{\mathcal{K}}(\mathbb{R}_+)$ is the space of square-integrable functions with values in some Hilbert space $\mathcal{K}$ [185]). In general the representation of such a process requires infinitely many independent creation, annihilation and number processes.

Among processes with independent increments only the Wiener and Poisson processes have the following property of *chaos decomposition*: the Hilbert space of square-integrable functionals of the process is the direct sum of subspaces generated by n-fold iterated stochastic integrals (Wiener, Ito). The question – which other martingales have this property has attracted attention of specialists in the theory of stochastic processes. In particular, Emery showed that this property is also possessed by the Azema martingale

$$X_t = \mathrm{sgn}W_t \sqrt{2(t - g_t)}, \tag{5.30}$$

where g_t is the last zero of the Wiener process W_t before time t. Parthasarathy [184] studied the quantum stochastic differential equation

$$dX(t) = (c-1)X(t)dA(t) + dQ(t), \quad X(0) = 0,$$

and showed that for arbitrary $c \in [-1,1]$ it has a solution comprising a commuting family of selfadjoint operators and is stochastically equivalent (in the vacuum state) to a martingale with the chaos decomposition property. As Meyer noted, when $c = 0$, $X(t)$ is stochastically equivalent to the Azema martingale. Thus, the highly non-linear transformation (5.30) of the Wiener process proves to be closely related to a linear stochastic differential equation for noncommuting processes.

5.2.2 Stochastic Evolutions and Dilations of Dynamical Semigroups

An interesting class of infinitely divisible objects arises in connection with dynamical semigroups. Let $\mathcal{H}$ a Hilbert space, Φ a dynamical map in the algebra $\mathfrak{B}(\mathcal{H})$, that is, a normal completely positive map, such that $\Phi[I] = I$. We call Φ *infinitely divisible* if for arbitrary $n = 1, 2, \ldots$, $\Phi = (\Phi_n)^n$, where Φ_n is a dynamical map. If $\{\Phi_t; t \in \mathbb{R}_+\}$ is a dynamical semigroup, then all the maps Φ_t are infinitely divisible since $\Phi_t = (\Phi_{\frac{t}{n}})^n$. On the other hand, if $\dim \mathcal{H} < \infty$ then any infinitely divisible dynamical map has the form $\Phi = \mathcal{E} \cdot e^{\mathcal{L}}$, where $\mathcal{E}$ is the expectation onto a subalgebra $\mathfrak{B} \subset \mathfrak{B}(\mathcal{H})$ and $\mathcal{L}$ is a completely dissipative map leaving the subalgebra $\mathfrak{B}$ invariant [60]. The map $\mathcal{E}$ plays a role similar to an idempotent divisor in the theory of infinitely divisible positive definite functions on groups [186]. If $\mathcal{E} = \mathrm{Id}$ then we may associate a quantum dynamical semigroup with Φ.

Hudson and Parthasarathy [138] constructed an embedding of a norm continuous quantum dynamical semigroup in the Fock space, which may be interpreted as a dilation to a Markov quantum stochastic process in the sense of Sect. 3.3.6, Chap. 1, (see the paper of Frigerio in [194].) For simplicity, we restrict ourselves to a description of the construction of [138] for a semigroup with an infinitesimal generator of form

$$\mathcal{L}[X] = i[H, X] + L^* X L - L^* L \cdot X, \qquad (5.31)$$

where $H, L \in \mathfrak{B}(\mathcal{H}), H^* = H$.

Proposition 5.2.1. *Let $\{\Phi_t; t \in \mathbb{R}_+\}$ be a quantum dynamical semigroup in $\mathfrak{B}(\mathcal{H})$ with the infinitesimal operator (5.31), then*

$$\Phi_t[X] = \mathcal{E}[V(t)^*(X \otimes I)V(t))], \quad X \in \mathfrak{B}(\mathcal{H}), \qquad (5.32)$$

where $\{V(t); t \in \mathbb{R}_+\}$ is the family of unitary operators in $\widetilde{\mathcal{H}} = \mathcal{H} \otimes \Gamma(L^2(\mathbb{R}_+))$, satisfying equation (5.26), and the map $\mathcal{E}_0 : \mathfrak{B}(\widetilde{\mathcal{H}}) \to \mathfrak{B}(\mathcal{H})$ is the vacuum conditional expectation defined by

$$\mathrm{Tr}\, S\mathcal{E}_0[Y] = \mathrm{Tr}(S \otimes \mid \psi_0\rangle\langle\psi_0 \mid)Y.$$

for an arbitrary density operator S in $\mathcal{H}$ and arbitrary $Y \in \mathfrak{B}(\widetilde{\mathcal{H}})$.

Proof. From (5.26) and the quantum Ito formula follows the quantum Langevin equation for $X(t)$

$$dX(t) = [X(t)^*, X(t)]dA(t) + [X(t), L(t)]dA^+(t) +$$
$$+ \{i[H(t), X(t)] + (L(t)^* X(t)L(t) - L(t)^* L(t) \cdot X(t))\}dt.$$

Taking the vacuum expectation and recalling the relations

$$dA(t)\psi_0 = 0, \quad d\Lambda(t)\psi_0 = 0, \tag{5.33}$$

which follow from (5.6), we see that the family of observables $\tilde{\Phi}_t[X] = \mathcal{E}_0[V(t)^*(X \otimes I)V(t)]$ in the algebra $\mathfrak{B}(\mathcal{H})$ satisfies the equation

$$d\tilde{\Phi}_t[X] = \tilde{\Phi}_t[\mathcal{L}[X]]dt; \quad \tilde{\Phi}_0[X] = X.$$

From this it follows, that

$$\tilde{\Phi}_t[X] = \exp t\mathcal{L}[X] = \Phi_t[X].$$

Now let $\{\psi_t; \ t \in \mathbb{R}_+\}$ be the corresponding dynamical semigroup in the space of states. From the representation (5.32) follows a constructive proof of the dilation theorem, formulated in Sect. 3.3.6 of Chap. 3. Denote by $\mathcal{H}_0 = \Gamma(\mathrm{L}^2(\mathbb{R}))$ the Fock space over $\mathrm{L}^2(\mathbb{R})$ and let $S_0 = |\psi_0\rangle\langle\psi_0|$ where ψ_0 is the vacuum vector in $\mathrm{L}^2(\mathbb{R})$. In the space $\mathcal{H} \otimes \mathcal{H}_0$ consider the group of unitary time-shift operators $\{S_t; \ t \in \mathbb{R}\}$, defined as in (5.4). Since $\mathcal{H}_0 = \Gamma(\mathrm{L}^2(\mathbb{R}_-)) \otimes \Gamma(\mathrm{L}^2(\mathbb{R}_+))$, where $\mathbb{R}_- = (-\infty, 0)$, the action of the fundamental processes $A(t), A^+(t), \Lambda(t); \ t \in \mathbb{R}_+$; transfers naturally to $\mathcal{H} \otimes \mathcal{H}_0$. The solution of equation 5.26 is then the family of unitary operators $\{V(t); \ t \in \mathbb{R}_+\}$ in $\mathcal{H} \otimes \mathcal{H}_0$, satisfying the cocycle relation (5.19). From this relation it follows that

$$U_t = \begin{cases} S_t V(t), & t \in \mathbb{R}_+ \\ V(-t)^* S_t, & t \in \mathbb{R}_- \end{cases}$$

is a group of unitary operators in $\mathcal{H} \otimes \mathcal{H}_0$. Since from (5.32) $S_t^*(X \otimes I)S_t = X \otimes I$, it follows that

$$\Psi_t[S] = \mathrm{Tr}_{\mathcal{H}_0} U_t(S \otimes S_0)U_t^*, \ t \in \mathbb{R}_+.$$

The dilation constructed here has a transparent physical interpretation. The group of the time shifts $\{S_t\}$ describes the dynamics of a quantum noise which acts as the environment of the system under investigation. Expressing the operators $V(t)$ as time-ordered exponential (5.25), it is possible to see that they describe the evolution of the system with the Hamiltonian H interacting with the environment by means of the singular Hamiltonian

$$H_{\text{int}} = i(L\dot{A}^+(t) - L^*\dot{A}(t)). \tag{5.34}$$

The averaging of the unitary evolution $\{U_t\}$ over the vacuum state gives a dynamical semigroup in $\mathfrak{B}(\mathcal{H})$.

A similar unitary dilation holds for an arbitrary quantum dynamical semigroup with infinitesimal operator (3.3.4) but one needs to use quantum stochastic calculus with an infinite number of creation and annihilation operators.

From the point of view of statistical mechanics it is interesting to elucidate the exact conditions under which such a highly idealized dynamical system as quantum noise arises from more realistic physical models of open systems (in this connection see [1], where the weak coupling and low density approximations are discussed.)

From (5.32) we may also obtain representations of a quantum dynamical semigroup by solutions of classical stochastic differential equations in the Hilbert space $\mathcal{H}$ [117]. Let W_t be a standard Wiener process and let $\{V_t^{(1)}(W); t \in \mathbb{R}_+\}$ be a stochastic process with values in $\mathfrak{B}(\mathcal{H})$, satisfying the stochastic differential equation

$$dV_t^{(1)}(W) = \left[LdW_t - \left(iH + \tfrac{1}{2}L^*L\right)dt\right]V_t^{(1)}(W); \atop V_0^{(1)}(W) = I \tag{5.35}$$

Then

$$\Phi_t[X] = \mathbf{E}_{(1)}(V_t^{(1)}(W)^* X V_t^{(1)}(W)), \tag{5.36}$$

where $\mathbf{E}_{(1)}(\cdot)$ is the expectation over the Wiener process. On the other hand, let N_t be a Poisson process of unit intensity and let $\{V_t^{(2)}(N); t \in \mathbb{R}_+\}$ satisfy the equation

$$dV_t^{(2)}(N) = \left[(L - I)dN_t - \left(iH + \tfrac{1}{2}(L^*L - 1)\right)dt\right]V_t^{(2)}(N); \atop V_0^{(2)}(N) = I. \tag{5.37}$$

Then

$$\Phi_t[X] = \mathbf{E}_{(2)}(V_t^{(2)}(N)^* X V_t^{(2)}(N)) \tag{5.38}$$

where $\mathbf{E}_{(2)}()$ is the expectation over the Poisson process.

Note that the solutions of the equations (5.35), (5.37) can be written as time-ordered exponentials (multiplicative stochastic integrals)

$$V_t^{(1)}(W) = \overleftarrow{\exp} \int_0^t \left\{ L dW_s - \left[iH + \frac{1}{2}(L^* + L)L \right] ds \right\}, \tag{5.39}$$

$$V_t^{(2)}(N) = \overleftarrow{\exp} \int_0^t \left\{ (\ln L) dN_s - \left[iH + \frac{1}{2}(L^*L - I) \right] ds \right\}, \tag{5.40}$$

We content ourselves with the derivation of (5.36). Introduce the family of isometric operators $V_t^{(1)}$ from $\mathcal{H}$ to $\mathfrak{H} = \mathcal{H} \otimes \Gamma(L^2(\mathbb{R}_+))$ defined by

$$V_t^{(t)}\psi = V(t)(\psi \otimes \psi_0); \quad \psi \in \mathcal{H}. \tag{5.41}$$

From (5.1.4) it follows that

$$dV_t^{(1)} = \left[L dQ(t) - \left(iH + \frac{1}{2}L^*L \right) dt \right] V_t^{(1)}, \tag{5.42}$$

since by (5.33), the coefficient of $dA(t)$ may be arbitrary. If we use the duality map then (5.32) becomes (5.36) and (5.42) becomes (5.35). Similarly, the representation (5.38) follows from (5.32) using the map $J^{(\lambda)}$ of (5.29) (for $\lambda = 1$, see (5.32)).

5.2.3 Dilations of Instrumental Processes

In the study of continuous measurement processes a concept of infinite divisible instrument arises naturally combining the infinite divisibility of probability distributions and that of dynamical maps. Barchielli and Lupieri [17], [194] constructed the corresponding dilation in the Fock space, which may be considered as a concrete realization of Ozawa's theorem from Sect. 4.1.2, Chap. 4, for measurement processes running continuously in time. We restrict ourselves here to the two most important examples.

Example 5.2.1. Consider the *i*-process $\{\mathcal{N}_{a,b}^{(1)}\}$ with the generator

$$\mathcal{L}^{(1)}(\lambda)[X] = \mathcal{L}[X] + i\lambda(L^*X + XL) - \frac{1}{2}\lambda^2 X, \tag{5.43}$$

where $L, H = H^*$ are bounded operators in the Hilbert space $\mathcal{H}$, and $\mathcal{L}[X]$ is given by (5.31). According to Sect. 4.2.3, Chap. 4, this is a continuous measurement process of the observable $A = (L + L^*)$ in a system evolving with the Hamiltonian H. It follows from the result of Barchielli and Lupieri [17], that

$$N_{0,t}^{(1)}(E)[X] = \mathcal{E}_0[(V(t)^*(X \otimes P_{0,t}^{(1)}(E))V(t)]; \quad E \in \mathcal{B}_{0,t}, \tag{5.44}$$

where $\{V(t)\}$ is the family of unitary operators in the Hilbert space $\mathfrak{H} = \mathcal{H} \otimes \Gamma(L^2(\mathbb{R}_+))$ satisfying the quantum stochastic differential equation (5.26), and $P_{0,t}^{(1)}(E)$; $E \in \mathcal{B}_{0,t}$, is the spectral measure of the family of compatible observables $Q(s)$; $0 \le s \le t$, in $\Gamma(L^2(\mathbb{R}_+))$, see Sect. 5.2.1 (because of homogeneity, a similar representation also holds for $N_{a,b}^{(1)}$, where $a \le b$).

Let $\{M_{a,b}^{(1)}\}$ be the corresponding i-process in the state space. Then the relation (5.44) takes the form of (4.10) from Chap. 4

$$M_{0,t}^{(1)}(E)[S] = \mathrm{Tr}_{\Gamma(L^2(\mathbb{R}_+))} V(t)(S \otimes |\psi_0\rangle\langle\psi_0|) \times$$
$$\times V(t)^*(I \otimes P_{0,t}^{(1)}(E)), \tag{5.45}$$

which has a clear physical interpretation: the observed system, being initially in the state S and evolving with the Hamiltonian H, interacts with the quantum noise by means of the singular Hamiltonian (5.34). A continuous nondemolition measurement of the family of compatible observables $Q(s)$; $0 \le s \le t$ is made over the quantum noise (which plays the part of the probe system).

Example 5.2.2. The i-process $\{N_{a,b}^{(2)}\}$ with the generator

$$\mathcal{L}^{(2)}(\lambda)[X] = i[H, X] + (L^* X L e^{i\lambda} - L^* L \cdot X), \tag{5.46}$$

where $H, L \in \mathcal{B}(\mathcal{H})$, is the natural generalization of the counting process of Sect. 4.2.5, Chap. 4. For this the dilation has the form

$$N_{0,t}^{(2)}(E) = \mathcal{E}_0[V(t)^*(I \otimes P_{0,t}^{(2)}(E))V(t)]; \quad E \in \mathcal{B}_{0,t}, \tag{5.47}$$

where $P_{0,t}^{(2)}(E)$; $E \in \mathcal{B}_{0,t}$, is the spectral measure of the family of compatible observables $\Lambda(s)$; $0 \le s \le t$, in $\Gamma(L^2(\mathbb{R}_+))$. The corresponding representation in the space of states

$$M_{0,t}^{(2)}(E) = \mathrm{Tr}_{\Gamma(L^2(\mathbb{R}_+))} V(t)(S \otimes |\psi_0\rangle\langle\psi_0|)V(t)^*(I \otimes P_{0,t}^{(2)}(E)) \tag{5.48}$$

has an interpretation similar to that of (5.45)

A few words on the method of proof of (5.44), (5.47). Denote

$$\tilde{\mathcal{N}}_t^{(j)}(B) = \mathcal{E}_0[V(t)^*(I \otimes P_{0,t}^{(j)}(Y(\cdot) : Y(t) - Y(0) \in B))V(t)],$$

where $B \in \mathcal{B}(\mathbb{R})$, and introduce the characteristic function

$$\tilde{\Phi}_t^{(j)}(\lambda) = \int_{\mathbb{R}} e^{i\lambda x} \tilde{\mathcal{N}}_t^{(j)}(dx).$$

Then

$$\tilde{\Phi}_t^{(j)}(\lambda)[X] = \mathcal{E}_0[V(t)^*(X \otimes e^{i\lambda Y^{(j)}(t)})V(t)], \qquad (5.49)$$

where $Y^{(1)}(t) = Q(t)$ and $Y^{(2)}(t) = \Lambda(t)$. Using the quantum Ito formula it can be seen that the function (5.49) satisfies the equations

$$d\tilde{\Phi}_t^{(j)}(\lambda) = \mathcal{L}^{(j)}(\lambda) \cdot \tilde{\Phi}_t^{(j)}(\lambda)dt,$$
$$\tilde{\Phi}_t^{(j)}(\lambda) = \exp t\mathcal{L}^{(j)}(\lambda).$$

From this it follows that $\{\tilde{\mathcal{N}}_t^{(j)}\}$ is the convolution semigroup corresponding to the i-process $\{\mathcal{N}_{a,b}^{(j)}\}$. By the bijectivity of the correspondence between the i-process and the convolution semigroups (see Sect. 4.2.5), (5.44) and (5.47) follow.

5.2.4 Stochastic Representations
of Continuous Measurement Processes

Using the trick which enabled us in (5.30) to obtain a stochastic representation of a quantum dynamical semigroup, we can give a corresponding stochastic representation for a continuous measurement process, see [118]. From these representations one can obtain an explicit description of the probability distribution of the outcomes of continuous measurement in the space of trajectories and of posterior states of the quantum system under observation.

Consider first the i-process with the generator (5.43), which as noted in Sect. 4.2.5, is concentrated on continuous trajectories. Let $\mu_{(1)}$ be the Wiener measure in the space of continuous functions $\mathcal{C}$, corresponding to the standard Wiener process W_t.

Proposition 5.2.2. *The i-process is absolutely continuous with respect to the measure $\mu_{(1)}$ in the sense that*

$$\mathcal{M}_{0,t}^{(1)}(E)[S] = \int_E V_t^1(W) S V_t^{(1)}(W)^* d\mu_{(1)}(W); \quad E \in \mathcal{B}_{0,t} \cap \mathcal{C}, \qquad (5.50)$$

where $\{V_t^{(1)}(W)\}$ is a family of bounded operators in $\mathcal{H}$ satisfying the stochastic differential equation (5.35).

The proof is based on the application of the duality transform to the representation (5.45). Here the operators $V_t^{(1)}(W)$ are as in Sect. 5.2.2 while the spectral measure $P_{0,t}^{(1)}$ diagonalizes the family $Q(s); 0 \le s \le t$, so that the projector $\Gamma_{0,t}^{(1)}(E)$ goes into the indicator of the set $E \in \mathcal{B}_{0,t} \cap \mathcal{C}$.

The relation (5.45) gives a concrete representation of a completely positive instrument $\mathcal{M}_{0,t}^{(1)}$ in the form (4.13). From this we obtain the probability distribution in the space of observed trajectories

$$\mu_S(E) = \int_E \mathrm{Tr}\, S V_t^{(1)}(W)^* V_t^{(1)}(W) d\mu_{(1)}(W); \quad E \in \mathcal{B}_{0,t} \cap \mathcal{C}.$$

It is absolutely continuous with respect to the Wiener measure $\mu_{(1)}$ and has the density

$$P_t^{(1)}(W) = \mathrm{Tr}\, S V_t^{(1)}(W) V_t^{(1)}(W)^*, \qquad (5.51)$$

which is almost surely positive. The posterior state, corresponding to the observed trajectory $W_s; 0 \le s \le t$, is

$$S_t^{(1)}(W) = P_t^{(1)}(W)^{-1} V_t^{(1)}(W) S V_t^{(1)}(W)^*. \qquad (5.52)$$

We note that if the initial state is pure $S = |\psi\rangle\langle\psi|$, then the posterior states are pure $S_t^{(1)}(W) = |\psi_t^{(1)}(W)\rangle\langle\psi_t^{(1)}(W)|$, where

$$\psi_t^{(1)}(W) = V_t^{(1)}(W)\psi / \|V_t^{(1)}(W)\psi\|.$$

Let us pass on to the case of the counting process with the generator (5.46). Let $\mu_{(2)}$ be the measure on the space $\mathcal{D}$ corresponding to the Poisson process of unit intensity.

Proposition 5.2.3. *The i-process $\{\mathcal{M}_{a,b}^{(2)}\}$ is absolutely continuous with respect to the measure $\mu_{(2)}$, furthermore*

$$M_{0,t}^{(2)}(E)[S] = \int_E V_t^{(2)} S V_t^{(2)}(N)^* d\mu_{(2)}(N); \quad E \in \mathcal{B}_{0,t} \cap \mathcal{D}, \qquad (5.53)$$

where $\{V_t^{(2)}(N)\}$ is a family of bounded operators in $\mathcal{H}$ satisfying the stochastic differential equation (5.37) $\mu_{(2)}$.

The proof of (5.33) requires a transformation of the representation (5.48). Consider the unitary Weyl operators $V_z(t) = \exp[zA^+(t) - zA(t)]$, where $z \in \mathbb{C}$. For $s \leq t$ the following relation holds

$$V_z(t)^* A(s) V_z(t) = A(s) + \bar{z}A(s) + zA^+(s) + |z|^2 s \equiv \Pi(s),$$

which can be verified using (5.27) and the quantum Ito formula. Let $z = 1$, then $\Pi(s)$ is a Poisson process of unit intensity in the Fock space. Setting $\tilde{U}_t = V_1(t)^* V(t)$, we rewrite (5.48) in the form

$$M_{0,t}^{(2)}(E) = \operatorname{Tr}_{\Gamma(\mathrm{L}^2(\mathbb{R}_+))} \tilde{U}_t (S \otimes |\psi_0\rangle\langle\psi_0|) \tilde{U}^* (I \otimes \tilde{P}_{0,t}^{(2)}(E)), \qquad (5.54)$$

where $\tilde{P}_{0,t}^{(2)}$ is a spectral measure of the family of compatible observables $\Pi(s)$; $0 \leq s \leq t$. From the quantum Ito formula follows the equation for $\tilde{U}_t$;

$$d\tilde{U}_t = \left\{ (L - I)dA^+(t) - (L - I)^* dA(t) - \left[iH + \frac{1}{2}(L^*L - 2L + I) \right] dt \right\} \tilde{U}_t.$$

Introducing the isometric operators $V_t^{(2)}$ from $\mathcal{H}$ into $\mathfrak{H}$ defined by

$$V_t^{(2)}\psi = \tilde{U}_t(\psi \otimes \psi_0); \quad \psi \in \mathcal{H},$$

and bearing in mind relation (5.33), we obtain

$$dV_t^{(2)} = \left\{ (L - 1)d\Pi(t) - \left[iH + \frac{1}{2}(L^*L - 2L - I) \right] dt \right\} V_t^{(2)}. \qquad (5.55)$$

The unitary operator $J^{(1)}$ from 5.2.1 takes the Fock space $\Gamma(\mathrm{L}^2(\mathbb{R}_+))$ into $\mathrm{L}^2(N)$, where N_t is a classical Poisson process of unit intensity, while (5.55) is transformed into (5.37). The projector $\tilde{P}_{0,t}^{(2)}(E)$ goes into the indicator of the set E, and (5.54) – into (5.53).

Formulas similar to (5.51)–(5.53) are obtained for probability distributions in the space of observable trajectories and posterior states.

5.2.5 Nonlinear Stochastic Equations of Posterior Dynamics

Here we obtain stochastic differential equations satisfied by the observable trajectories and posterior states in a continuous quantum measurement process. Consider first the process of measurement of the observable $A = L + L^*$

with the generator (5.43). From equation (5.35) for the family $V_t^{(1)}(W)$ and from (5.51) it follows that the density $P_t^{(1)}(W)$ of the probability distribution of observables of the trajectories M_s with respect to the Wiener measure $\mu_{(1)}$ satisfies the stochastic differential equation

$$dp_t^{(1)}(W) = m_t(W)p_t^{(1)}(W)dW_t, \qquad (5.56)$$

where

$$m_t(W) = \mathrm{Tr} S_t^{(1)}(W)A$$

is the posterior mean of the observable A. From this it follows that the observed process $Y(t)$ is a diffusion type process satisfying stochastic differential equation

$$dY(t) = m_t(Y)dt + d\tilde{W}_t, \qquad (5.57)$$

where $\tilde{W}_t$ is the innovating Wiener process[2]. By applying Ito stochastic calculus one obtains equation for the posterior state (5.52)

$$dS_t^{(1)}(Y) - \mathcal{K}[S_t^{(1)}(Y)]dt = [(L - \frac{1}{2}m_t(Y))S_t^{(1)}(Y) +$$
$$+ S_t^{(1)}(Y)(L - \frac{1}{2}m_t(Y))^*][dY(t) - m_t(Y)dt], \qquad (5.58)$$

where

$$\mathcal{K}[S] = -i[H, S] + LSL^* - L^*L \circ S.$$

This equation was proposed by Belavkin [26] (see also Diosi [61]) in studying the quantum analogue of the filtering problem for stochastic processes. In the case of pure states the equation for the vector of the posterior state has the form

$$d\psi_t^{(1)}(Y) = [L - \frac{1}{2}m_t(Y)]\psi_t^{(1)}(Y)[dY(t) - m_t(Y)dt] -$$
$$- \left[iH + \frac{1}{2}(L^*L - Lm_t(Y) + \frac{1}{4}m_t(Y)^2\right]\psi_t^{(1)}(Y)dt. \quad (5.59)$$

---

[2] See R.S. Liptser and A.N. Shiryaev, The Statistics of Stochastic Process, Nauka, Moscow, (1974), Sect. 7.4.

The nonlinearity of the equations (5.58), (5.59) is caused by the normalization of the posterior states (5.52), while the underlying stochastic equation (5.35), which is an analog of Zakai's equation in classical filtering theory, is linear.

Of great interest is the problem of the derivation and study of equations (5.58), (5.59) in the case when L, H are unbounded operators. Diosi [61], Belavkin and Staszewski [29], Belavkin and Melsheimer [28] considered the equation

$$d\psi_t^{(1)}(Y) = (Q - <Q>_t)\psi_t^{(1)}(Y)[dY(t) - 2 <Q>_t dt] -$$
$$- \left[iP^2/2m + \frac{1}{2}(Q - <Q>_t)^2 \right] \psi_t^{(1)}(Y)dt, \qquad (5.60)$$

which is obtained from (5.59) by the formal substitutions $L = Q$, $H = p^2/2m$, where P, Q are canonical observables of a non-relativistic particle of mass m. This corresponds to a continuous approximate measurement of the position of a free particle. An explicit solution is found in the case of Gaussian initial state. It is shown that it is Gaussian with the variance having a finite limit as $t \to +\infty$. Thus equation (5.60) resolves the well-known quantum mechanical paradox with spreading of the wave packet of a free particle.

It is interesting that similar nonlinear equations, but with a completely different motivation and interpretation, arose almost simultaneously in the papers of authors engaged in a search for an alternative conceptual foundation for the quantum measurement theory. In a paper published in [194] Ghirardi, Rimini and Weber posed a problem of finding equations giving a unified description of micro- and macro-systems, from which, in particular, there would follow both reversible quantum dynamics and irreversible changes of the projection postulate type. Different solutions of this problem have been proposed; in the papers of Gisin [81], [78], an equation of the type (5.59) was introduced in which, however, $dY(t) - m_t(Y)dt$ is replaced by a stochastic differential of a certain a priori given Wiener process (the equation in [81] is distinguished by the choice of the phase factor for $\psi_t^{(1)}$). Let $H = 0, L = \sum x_i E_i$ be a selfadjoint operator with a purely point spectrum. In [81], [78] it is noted that the equation obtained

$$d\psi_t = (L - \langle L \rangle_t)\psi_t dW_t - \frac{1}{2}(L - \langle L \rangle_t)^2 \psi_t dt, \qquad (5.61)$$

where $\langle L \rangle_t = \langle \psi_t | L | \psi_t \rangle$, gives a dynamical description of the projection postulate $\psi \to \psi_i = E_i\psi/\|E_i\psi\|$, in the sense that as $t \to +\infty$ the solution ψ_t tends to one of the states ψ_i. Gatarek and Gisin [76] analyzed equation (5.61) for an unbounded operator L and also equations of type (5.60). To prove the existence of a weak solution these authors used a method of change of probability measure (the Girsanov transformation), which in the continuous measurement scheme has the meaning of a transition from the process $Y(t)$ to the Wiener process $\hat{W}_t$, defined by (5.57). They also commented that the multidimensional case can be treated in the same way. However the argument in fact substantially relies upon the property (5.65) below, which is automatic in the case of (5.61). To extend this argument to the general case, one has to complement it by a condition, ensuring the mean-square norm conservation (see below).

In the case of the counting process with generator (5.46), the stochastic differential equation for the density in the space of trajectories has the form

$$dp_t^{(2)}(N) = [\lambda_t(N) - 1]p_t^{(2)}(N)(dN - dt),$$

where

$$\lambda_t(N) = \operatorname{Tr} S_t^{(2)}(N)L^*L$$

is the posterior intensity of jumps. The equation for the posterior state is

$$dS_t^{(2)}(Y) - \mathcal{K}[S_t^{(2)}(Y)]dt \tag{5.62}$$
$$= \left[\frac{LS_t^{(2)}(Y)L^*}{\lambda_t(Y)} - I\right][dY(t) - \lambda_t(Y)dt].$$

The general form of *dissipative* Ito stochastic integro–differential equations reducing to the nonlinear equations of the type (5.58) or (5.62) in the case of Wiener or Poisson noises, and the related quantum processes were studied by Barchielli and Holevo [19]. In the papers of Holevo [126], [128], the dissipative stochastic equations driven by multidimensional Wiener process were studied in the case of the general unbounded operator coefficients and applied to construction of quantum dynamical semigroups with unbounded generators.

The classical stochastic calculus provides a powerful analytic tool for the study of related quantum processes; on the other hand, quantum stochastic calculus gives a new insight into the existence and uniqueness problems for a broad class of the Ito stochastic equations in the Hilbert space. Consider the linear stochastic equation

$$\psi_t = \psi_0 + \int_0^t \sum_j L_j \psi(s)dW_j(s) - \int_0^t K\psi(s)ds \tag{5.63}$$

in a separable Hilbert space $\mathcal{H}$, where $W_j(t), j = 1, 2, \ldots$ are independent standard Wiener processes, L_j, K are, in general, unbounded operators defined on a dense domain $\mathcal{D} \subset \mathcal{H}$ and $\psi_0 \in \mathcal{H}$. The equation is called *dissipative* if

$$\sum_j \|L_j\psi\|^2 - 2\Re\langle K\psi|\psi\rangle \le c\|\psi\|^2, \ \psi \in \mathcal{D}. \tag{5.64}$$

Results on the equation of this type can be found in the book of Rozovsky [200], where existence and uniqueness of the strong solution of the equation (5.63) in a scale of Hilbert spaces is established under additional hypotheses. Many other works use conditions of the coercivity type, stronger than (5.64). Under very mild conditions it is possible to prove the existence and uniqueness for generalized solutions of dissipative equations in the weak topology sense [126]. The proof takes some inspirations from quantum stochastic calculus (although not using it!), namely, from Frigerio-Fagnola's existence proof and Mohari's uniqueness proof for quantum stochastic differential equations (see [170], Ch. VI, for a survey of these ideas). The dual stochastic differential equation can be introduced, whose solution is a classical analog of the "dual cocycle", introduced and studied in the noncommutative situation by Journé.

The conservative stochastic differential equations appear to be closely related to important concepts in noncommutative probability, such as dynamical semigroups and the nonlinear stochastic equation for the normalized posterior wave function. The linear equation (5.63) is *conservative* if the left-hand side in (5.64) is identically zero, i.e. (3.23) of Chap. 3 holds. If the operators L_j, K are bounded, then this condition implies mean-square norm conservation for the solution:

$$\mathsf{M}\|\psi_t\|^2 = \|\psi_0\|^2, \tag{5.65}$$

however in general one will have only $\mathsf{M}\|\psi_t\|^2 \le \|\psi_0\|^2$ with the possibility of strict inequality. Solutions of (5.63) provide classical stochastic representation for a quantum dynamical semigroup generalizing relation (5.36), and the property (5.65) is closely related to the unitality (non-explosion) of this dynamical semigroup. The equation (5.63) is related to the forward, while the dual equation is similarly related to the backward Markov master equation for which the semigroup is the common minimal solution. By using techniques of Ito stochastic calculus in a Hilbert space, (5.65) can be derived from conservativity and further condition of hyperdissipativity, which means dissipativity in a Hilbert scale associated to some strictly positive self-adjoint operator (see Sect. 3.3.3 of Chap. 3).

Bibliography

1. L. Accardi, R. Alicki, A. Frigerio, Y.G. Lu, *An invitation to the weak coupling and low density limits*, Quant. Prob. Rel. Topics **6**, 3-62 (1991)
2. L. Accardi, A. Frigerio, J. T. Lewis, *Quantum stochastic processes*, Publ. RIMS Kyoto University **18**, 97-133 (1982)
3. L. Accardi, Y.G. Lu, I. V. Volovich, *Stochastic limit of quantum theory*, Springer, Berlin, (2001) (to be published)
4. L. Accardi, M. Schürmann and W. von Waldenfels, *Quantum independent increment processes on superalgebras*, Math. Z. **198**, 451-477 (1988)
5. C. Adami, N. J. Cerf, *Capacity of noisy quantum channels*, Phys. Rev. A **56**, 3470-3485 (1997)
6. G. S. Agarwal, *Quantum theory of nonlinear mixing in multimode fields*, Phys. Rev. A A **34**, 4055-4069 (1986)
7. N.I. Akhiezer and I.M. Glazman, *The theory of linear operators in Hilbert space*, Moscow, Nauka, (1966)
8. P. M. Alberti, A. Uhlmann, *Stochasticity and partial order*, Kluwer, Dordrecht, (1982)
9. S.T. Ali and E. Prugovečki, *Mathematical problems of stochastic quantum mechanics: harmonic analysis on phase space and quantum geometry*, Acta Appl. Math. **6**, 1-62 (1986)
10. R. Alicki and K. Lendi, *Quantum dynamical semigroups and applications*, Lect. Notes Phys. **286**, (1987)
11. G. G. Amosov, A. S. Holevo, R. F. Werner, *On some additivity problems in quantum information theory*, Probl. of Inform. Transm. **36** No. 4, 25-34 (2000); LANL e-print math-ph/0003002
12. T. Ando and M.D. Choi, *Non-linear completely positive maps*. In *Aspects of positivity in functional analysis*. Eds. R. Nagel, U. Schlotterbeek and M.P.H. Wolff, North-Holland, Elsevier, pp. 3-13, (1986)
13. H. Araki, *Factorizable representations of current algebras*, Publ. RIMS, Kyoto Univ. Ser. **A5**, 361-422 (1970)
14. A.Yu. Artem'ev, *The classification of quantum Markov kinetic spin systems by symmetry groups of the surroundings*, Teoret. Mat. Fiz. **79** No. 3, 323-333 (1989)
15. A. Barchielli, L. Lanz and G.M. Prosperi, *A model for macroscopic description and continuous observations in quantum mechanics*, Nuovo Cimento **72 B**, 79-91 (1982)
16. A. Barchielli, L. Lanz and G.M. Prosperi, *Statistics of continuous trajectories in quantum mechanics: operation-valued stochastic processes*, Found. Phys. **13**, 779-812 (1983)
17. A. Barchielli and G. Lupieri, *Quantum stochastic calculus, operation-valued stochastic processes and continual measurements in quantum mechanics*, J. Math. Phys. **26** No. 9, 2222-2230 (1985)

18. A. Barchielli, G. Lupieri, *An analog of Hunt's representation theorem in quantum probability*, J. Theor. Probab. **6**, 231-265 (1993)
19. A. Barchielli, A. S. Holevo, *Constructing quantum measurement processes via classical stochastic calculus*, Stoch. Proc. and Appl. **58**, 293-317 (1995)
20. A. Barchielli, A. M. Paganoni, *Detection theory in quantum optics: stochastic representation*, Quant. Semiclass. Optics **8**, 133-156 (1996)
21. C. Barnett, R.E. Streater and I.E. Wilde, *The Ito-Clifford integral*, J. Funct. Anal. **48** No. 2, 171-212 (1982)
22. H. Barnum, M. A. Nielsen, B. Schumacher, *Information transmission through a noisy quantum channel*, Phys. Rev. A **57**, 4153-4175 (1998)
23. H. Barnum, E. Knill, M. A. Nielsen, *On quantum fidelities and channel capacities*, IEEE Trans. Inform Theory **84**, 1317-1329 (2000); LANL Report quant-ph/9809010
24. V.P. Belavkin, *The reconstruction theorem for a quantum random process*, Teoret. Mat. Fiz. **62** No. 3, 409-431 (1985)
25. V.P. Belavkin, *Quantum branching processes and non-linear dynamics of multiple quantum systems*, Dokl. Akad. Nauk SSSR **301** No. 6, 1348-1352, (1988)
26. V.P. Belavkin, *Nondemolition stochastic calculus in Fock space and nonlinear filtering and control in quantum systems*. In: *Stochastic methods in mathematics and physics, Proc. 24 Karpacz winter school 1988*. Eds. R. Gielerak and W. Karwowski, World Scientific, Singapore, pp. 310-324, (1989)
27. V.P. Belavkin and B.A. Grishanin, *A study of the problem of optimal estimation in quantum channels by the generalized Heisenberg inequality method*, Problemy Peredachi Informatsii **9** No. 3, 44-52 (1973)
28. V.P. Belavkin and O.Melsheimer, *A Hamiltonian solution to quantum collapse, state diffusion and spontaneous localization*. In: *Proc. 2-nd Int. Conf. on Quantum Communication and Measurement, Nottingham 1994*. Eds. V. P. Belavkin, O. Hirota and R. L. Hudson, Plenum Press, NY pp. 201-222, (1995)
29. V.P. Belavkin and P. Staszewski, *A quantum particle undergoing continuous observation*, Phys. Lett. A. **140** No. 7/8, 359-362 (1989)
30. J.S. Bell, *On the problem of hidden variables in quantum mechanics*, Rev. Modern Phys. **38**, 447-552 (1966)
31. C. Benjaballah, *Introduction to photon communication*, Lect. Notes Phys. **m29**, Springer-Verlag, Berlin, (1995)
32. C. H. Bennett, P. W. Shor, *Quantum information theory*, IEEE Trans. Inform. Theory **44** No. 6, 2724-2742 (1998)
33. C. H. Bennett, P. W. Shor, J. A. Smolin, A. V. Thapliyal, *Entanglement-assisted classical capacity of noisy quantum channel*, Phys. Rev. Lett., **83**, 3081, (1999); LANL Report quant-ph/9904023
34. S.K. Berberian, *Notes on spectral theory*, Van Nostrand, Princeton, (1966)
35. F.A. Berezin, *The method of second quantization*, Moscow, Nauka, (1986)
36. N.N. Bogoliubov, A.A. Logunov, A.I. Oksak and I.T. Todorov, *The general principles of quantum field theory*, Moscow, Nauka, (1987)
37. N.A. Bogomolov, *Minimax measurements in the general theory of statistical solutions*, Teor. Veroyatnost. i Primen. **26** No. 4, 798-807 (1981)
38. V. B. Braginsky, F. Ya. Khalili, *Quantum measurement*, Cambridge University Press, Cambridge, (1992)
39. V.B. Braginsky, Y.I. Vorontzov and K.S. Thorne, *Quantum nondemolition measurement*, Science **209** 4456, 547-557 (1980)
40. O. Bratteli and D.W. Robinson, *Operator algebras and quantum statistical mechanics I*, Springer-Verlag, New York-Heidelberg-Berlin, (1979)
41. O. Bratteli and D.W. Robinson, *Operator algebras and quantum statistical mechanics II*, Springer-Verlag, New York-Heidelberg-Berlin, 1981

42. P. Busch, M. Grabovski, P. J. Lahti, *Operational quantum physics*, Lect. Notes Phys. m31, Springer-Verlag, New York-Heidelberg-Berlin, (1997)

43. H. J. Carmichael, *An open system approach to quantum optics*, Lect. Notes. Phys. m18, Springer-Verlag, New York-Heidelberg-Berlin, (1993)

44. P. Carruthers and M.M. Nieto, *Phase and angle variables in quantum mechanics*, Rev. Mod. Phys. 40 No. 2, 411-440 (1968)

45. D.P.L. Castrigiano, *On euclidean systems of covariance for massless particles*, Lett. Math. Phys. 5, 303-309 (1981)

46. U. Cattaneo, *On Mackey's imprimitivity theorem*, Commun. Math. Helv. 54 No. 4, 629-641 (1979)

47. U. Cattaneo, *Densities of covariant observables*, J. Math. Phys. 23 No. 4, 659-664 (1982)

48. A.M. Chebotarev, *Lectures on quantum probability*, Aportaciones matemáticas, textos 14, Sociedad Matemática Mexicana (2000)

49. A.M. Chebotarev, F. Fagnola, A. Frigerio, *Towards a stochastic Stone's theorem*, Stochastic partial differential equations and applications. Pitman Research Notes in Math. pp. 86-97, (1992)

50. A.M. Chebotarev, F. Fagnola, *Sufficient conditions for conservativity of quantum dynamical semigroups.*, J. Funct. Anal. 118 131-153 (1993)

51. N.N. Chentsov, *Statistical decision rules and optimal observations*, Moscow Nauka, (1972)

52. E. Christensen, D.E. Evans, *Cohomology of operator algebras and quantum dynamical semigroups*, J. London Math. Soc. 20 No. 2, 358-368 (1979)

53. *Contemporary problems of mathematics, The most recent achievements*, VINI-TI 36, (1990). English translation in J. Soviet Math. 56 No. 5, (1991)

54. T. M. Cover, J. A. Thomas, *Elements of information theory*, Wiley, New York, (1991)

55. G. M. D'Ariano, *Homodyning as universal detection*. In: *Quantum Communication, Computing and Measurement*. Eds. O. Hirota, A. S. Holevo and C. M. Caves, Plenum Press, NY, pp. 253-264, (1997); LANL e-print quant-ph/9701011

56. E.B. Davies, *Quantum theory of open systems*, Academic Press, London, (1976)

57. E.B. Davies, *Quantum dynamical semigroups ant the neutron diffusion equation*, Rep. Math. Phys. 11 No. 2, 169-188 (1977)

58. E.B. Davies, *Information and quantum measurement*, IEEE Trans. Inform. Theory IT 24, 596-599 (1978)

59. E. B. Davies and J.T. Lewis, *An operational approach to quantum probability*, Comm. Math. Phys. 17, 239-260 (1970)

60. L.V. Denisov, *Infinitely divisible Markov maps in quantum probability theory*, Teor. Veroyatnost. i Primenen. 33 No. 2, 417-420 (1988)

61. L. Diosi, *Continuous quantum measurement and Ito formalism*, Phys. Lett. A 129 No. 8/9, 419-423 (1988)

62. L. Diosi, *Localized solution of a simple nonlinear quantum Langevin equation*, Phys. Lett. A 132 No. 5, 233-236 (1988)

63. P.A.M. Dirac, *The principles of quantum mechanics*, 4th ed., Oxford Univ. Press, Oxford, (1958)

64. V.V. Dodonov and V.I. Manko, *Generalizations of uncertainty relations in quantum mechanics*, Trudy FIAN SSSR 183, 5-70 (1987)

65. H. D. Doebner, W. Luecke, *Quantum logic as a consequence of realistic measurements*, J. Math. Phys. 32, 250-253 (1991)

66. G.G. Emch, *Algebraic methods in statistical mechanics and quantum field theory*, Wiley Interscience, New York, (1972)

67. M. Emery, *Stabilité des solutions des équations differentielles stochastiques: applications aux integrals multiplicatives stochastiques*, Z. Wahrsch Verw. Gebiete **41**, 241-262 (1978)

68. D.E. Evans, *A review on semigroups of completely positive maps*, Lect. Notes Phys. **116**, 400-406 (1980)

69. D.E. Evans, J. T. Lewis, *Dilations on irreversible evolutions in algebraic quantum theory*, Comm Dublin Inst. Adv. Stud., Ser. A. **24**, (1977)

70. F. Fagnola, *Characterization of isometric and unitary weakly differentiable cocycles in Fock space*, Quant. Probab. Rel. Topics **8**, 189-214 (1993)

71. *Foundations of quantum mechanics and ordered linear spaces*. Eds. A. Hartkamper and H. Neumann, Lect. Notes Phys. **29**, Springer-Verlag, New York-Heidelberg-Berlin, (1974)

72. C.N. Friedman, *Semigroup product formulas, compressions and continual observations in quantum mechanics*, Indiana Univ. Math. J. **21** No. 11, 1001-1013 (1972)

73. A. Frigerio and H. Maassen, *Quantum Poisson processes and dilations of dynamical semigroups*, Probab. Theory Rel. Fields **83** No. 4, 489-508 (1989)

74. C. W. Gardiner, *Quantum noise*, Springer-Verlag, Berlin, (1991)

75. C. W. Gardiner, M. J. Collett, *Input and output in damped quantum systems: quantum stochastic differential equations and the master equation*, Phys. Rev. A **31**, 3761-3774 (1985)

76. D. Gatarek and N. Gisin, *Continuous quantum jumps and infinite dimensional stochastic equations*, J. Math. Phys. **32** No. 8, 2152-2157 (1991)

77. M. Gell-Mann, J. B. Hartle, *Quantum mechanics in the light of quantum cosmology*. In: *Proc. 3d Symp. Found. Quant. Mech.*. Tokyo, pp. 321-343, (1989)

78. G. Ghirardi, P. Pearle and A. Rimini, *Markov processes in Hilbert space and continuous spontaneous localization of systems of identical particles*, Preprint ICTP, Trieste IC/89/44, (1989)

79. R. D. Gill, S. Massar, *State estimation for large ensembles*, Phys. Rev. A **61**, 042312/1-16 (2000); LANL e-print quant-ph/9902063.

80. N. Giri and W. von Waldenfels, *An algebraic version of the central limit theorem*, Z. Warsch. Verw. Gebiete **42** No. 2, 129-134 (1978)

81. N. Gisin, *Stochastic quantum dynamics and relativity*, Helv. Phys. Acta **62**, 363-371 (1989)

82. R. Glauber, *The quantum theory of optical coherence*, Phys. Rev. **130**, 2529-2539 (1963)

83. J. P. Gordon, *Noise at optical frequencies; information theory*. In: *Quantum Electronics and Coherent Light, Proc. Int. School Phys. "Enrico Fermi", Course XXXI*. Ed. P. A. Miles, Academic Press, New York, pp.156-181 (1964)

84. V. Gorini, A. Frigerio, M. Verri, A. Kossakowski and E.C.G. Sudarshan, *Properties of quantum Markovian master equations*, Rep. Math. Phys. **13** No. 2, 149-173 (1978)

85. U. Grenander, *Probabilities on algebraic structures*, Almquist and Wiksell, Stockholm-Goteborg-Upsala, (1963)

86. A.A. Grib, *Weyl inequalities and the experimental verification of quantum correlations for macroscopic distances*, Uspekhi Fiz. Nauk **142** No. 4, 619-634 (1984)

87. U. Groh, *Positive semigroups on C^*- and W^*-algebras*, Lect. Notes Math. **1184**, 369-425 (1986)

88. S. Gudder, *Stochastic methods in quantum mechanics*, North-Holland, New York, (1979)

89. A. Guichardet, *Symmetric Hilbert spaces and related topics*, Lect. Notes Math. **261**, (1972)

90. R. Haag, D. Kastler, *An algebraic approach to quantum field theory*, J. Math. Phys. **5**, 848-861 (1964)

91. P. Hausladen, R. Jozsa, B. Schumacher, M. Westmoreland, W. Wootters, *Classical information capacity of a quantum channel*, Phys. Rev. A **54** No. 3, 1869-1876 (1996)

92. M. Hayashi. *Asymptotic estimation theory for a finite dimensional pure state model*, J. Phys. A **31**, 4633-4655, (1998)

93. G.C. Hegerfeldt, *Noncommutative analogs of probabilistic notions and results*, J. Funct. Anal. **64** No. 3, 436-456 (1985)

94. C.W. Helstrom, *Quantum detection and estimation theory*, Acad. Press, New York, (1976)

95. A.S. Holevo, *An analogue of the theory of statistical decisions in noncommutative probability*, Trudy Moscov Mat. Obshch **26**, 133-149 (1972)

96. A.S. Holevo, *Towards mathematical theory of quantum communication channels*, Probl. Pered. Inform. **8** No. 1, 62-71, (1972)

97. A.S. Holevo, *Information aspects of quantum measurement*, Probl. Pered. Inform. **9** No. 2, 31-42 (1973)

98. A.S. Holevo, *Some estimates for the information content transmitted by a quantum communication channel*, Probl. Pered. Inform. **9** No. 3, 3-11 (1973)

99. A.S. Holevo, *Some statistical problems for quantum Gaussian states*, IEEE Trans. Inform. Theory **21** No. 5, 533-543 (1975)

100. A.S. Holevo, *Investigations of a general theory of statistical decisions*, Trudy MIAN SSSR **124**, Moscow Nauka, (1976) . English translation: Proc. Steklov Inst. Math., No. 3 (1973)

101. A.S. Holevo, *Problems in the mathematical theory of quantum communication channels*, Rep. Math. Phys. **12** No. 2, 253-258 (1977)

102. A.S. Holevo, *Estimation of shift parameter of a quantum state*, Rep. Math. Phys. **13** No. 3, 287-307 (1978)

103. A.S. Holevo, *On the capacity of a quantum communication channel*, Probl. Pered. Inform. **15** No. 4, 3-11 (1979)

104. A.S. Holevo, *Covariant measurements and uncertainty relations*, Rep. Math. Phys.**16**, N3, 289-304 (1980)

105. A.S. Holevo, *Probabilistic and statistical aspects of quantum theory*, Moscow, Nauka, (1980). English translation: North Holland, Amsterdam, (1982)

106. A.S. Holevo, *On testing of statistical hypotheses in quantum theory*, Probab. Math. Statist. **3** No. 1, 113-126 (1982)

107. A.S. Holevo, *Quantum-probabilistic study of the counting statistics with application to the Dolinar receiver* Izv. VUZ. Matematika **26**, N8, 3-19 (1982). English translation: J. Soviet Math., 1-20

108. A.S. Holevo, *Bounds for generalized uncertainty of the shift parameter*, Lect. Notes Math. **1021**, 243-251 (1983)

109. A.S. Holevo, *Generalized imprimitivity systems for Abelian groups* Izv VUZ. Matematika, N2, 49-71 (1983). English translation: J. Soviet Math., 53-80

110. A.S. Holevo, *Covariant measurements and imprimitivity systems* Lect. Notes Math. **1055**, 153-172 (1984)

111. A.S. Holevo, *The statistical structure of quantum mechanics and hidden variables*, Moscow Znanie, 32 pp (1985)

112. A.S. Holevo, *Statistical definition of observables and the structure of statistical model*, Rep. Math. Phys. **22** No. 3, 385-407 (1985)

113. A.S. Holevo, *On a generalization of canonical quantization*, Izv. Akad. Nauk SSSR Ser. Mat. **50** No. 1, 181-194 (1986). English translation: Math. USSR Izvestiya Vol. **28** No. 1, 175-188 (1987)

114. A.S. Holevo, *Infinitely divisible measurements in quantum probability theory*, Teor. Veroyatnost. i Primen. **31** No. 3, 560-564 (1986)

115. A.S. Holevo, *Conditionally positive definite functions in quantum probability*. In: Proc. of International Congress of Mathematicians, Berkeley, Calif., USA, pp. 1011-1020 (1987)

116. A.S. Holevo, *Quantum estimation*, Adv. Statist. Signal Processing **1**, 157-202 (1987)

117. A.S. Holevo, *Stochastic representation of quantum dynamical semigroups*, Proc. Steklov Math. Inst. **191**, 130-139 (1989)

118. A.S. Holevo, *Inference for quantum processes*. In: *Proc. Internat. Workshop on Quantum Aspects of Optical Communication, Paris, 1990*. Eds. C. Benjaballah, O. Hirota and S. Reynaud, Lect. Notes in Phys. **378**, pp. 127-137 (1991)

119. A. S. Holevo, *Quantum probability and quantum statistics*, Itogi Nauki i Tehniki, ser. Sovr. Probl. Mat., **83** , VINITI, Moscow, (1991)

120. A.S. Holevo, *Time-ordered exponentials in quantum stochastic calculus*, Quant. Probab. Rel. Topics. **7**, 175-202 (1992)

121. A.S. Holevo, *A note on covariant dynamical semigroups*, Rep.Math.Phys. **32**, 211-216 (1993)

122. A.S. Holevo, *On conservativity of covariant dynamical semigroups*, Rep.Math.Phys. **33**, 95-110 (1993)

123. A.S. Holevo, *On the Levy-Khinchin formula in the non-commutative probability theory*, Theor. Probab. and Appl. **37**, N2, 211-216 (1993)

124. A.S. Holevo, *On the structure of covariant dynamical semigroups*, J. Funct. Anal. **131**, 255-278 (1995)

125. A.S. Holevo, *Excessive maps, "arrival times" and perturbations of dynamical semigroups*, Izvestiya: Mathematics., **59**, 207-222 (1995). English translation: 1311-1325.

126. A.S. Holevo, *On dissipative stochastic equations in Hilbert space*, Probab. Theory Rel. Fields **104**, 483-500 (1996)

127. A.S. Holevo, *Exponential formulae in quantum stochastic calculus*. Proc. Royal Academy of Edinburgh **126A**, 375-389 (1996)

128. A.S. Holevo, *Covariant quantum Markovian evolutions*, J. Math. Phys., **37** 1812-1832 (1996)

129. A.S. Holevo, *Stochastic differential equations in Hilbert space and quantum Markovian evolutions*, In: *Probability Theory and Mathematical Statistics. Proc. 7th Japan- Russia Symp. Tokyo, 1995*. Eds. S. Watanabe, M. Fukushima, Yu. V. Prohorov and A. N. Shiryaev, World Scientific, pp. 122-131 (1996)

130. A.S. Holevo, *There exists a non-standard dynamical semigroup on $\mathcal{B}(\mathcal{H})$*, Uspekhi Mat. Nauk., **51** No. 6, 225-226 (1996)

131. A.S. Holevo, *The capacity of quantum channel with general signal states*, IEEE Trans. Inform. Theory, **44** No. 1, 269-273 (1998); LANL e-print quant-ph/9611023,

132. A.S. Holevo, *Radon-Nikodym derivatives of quantum instruments*, J. Math. Phys., **39** No. 3, 1373-1387 (1998)

133. A.S. Holevo, *Quantum coding theorems*, Uspekhi Mat. Nauk, **53** No. 6, 193-230 (1998). English translation: Russian Math. Surveys, **53**, 1295-1331 (1998). LANL e-print quant-ph/9808023

134. A.S. Holevo, M. Sohma and O. Hirota, *The capacity of quantum Gaussian channels*, Phys. Rev. A **59**, 1820-1828, (1999).

135. A. S. Holevo, R. F. Werner, *Evaluating capacities of Bosonic Gaussian channels*, Phys. Rev. A **63**, 032312/1-18 (2001); LANL e-print quant-ph/9912067

136. R.L. Hudson, *Unification of Fermion and Boson stochastic calculus*, Comm. Math. Phys. **104**, 457-470 (1986)

137. R.L. Hudson and D. Applebaum, *Fermion Ito's formula and stochastic evolutions*, Comm. Math. Phys. 96, 456-473 (1984)

138. R.L. Hudson and K.R. Parthasarathy, *Quantum Ito's formula and stochastic evolutions*, Comm. Math. Phys. 93 No. 3, 301-323 (1984)

139. K.S. Ingarden, *Quantum information theory*, Rep. Math. Phys. 10 No. 1, 43-72 (1976)

140. P. Jajte, *Strong limit theorems in noncommutative probability*, Lect. Notes Math. 1110, (1985)

141. P.T. Jørgensen and R.T. Moore, *Operator commutation relations*, Reidel, Dordrecht, (1984)

142. T.L. Journé, *Structure des cocycles markoviens sur i espace de Fock*, Probab. Theory Rel. Fields 75 No. 2, 291-316 (1987)

143. R. Josza, B. Schumacher, *A new proof of the quantum noiseless coding theorem*, J. Modern Optics 41, 2343-2349 (1994)

144. L.A. Khalfin, B.S. Tsirelson, *Quantum and quasilocal analogs of Bell inequalities*. In: *Symp. on the foundations of modern physics*. Eds. P. Lahti and P. Mittelstaedt, pp. 441-460 (1985)

145. A. Yu. Kitaev, *Quantum computing: algorithms and error correction*, Uspekhi Mat. Nauk 52 No. 6, 53-112 (1997)

146. J.R. Klauder and E.C.G. Sudarshan, *Fundamentals of quantum optics*, W.A. Benjamin Inc., New-York-Amsterdam, (1968)

147. S. Kochen and E. Specker, *The problem of hidden variables in quantum mechanical systems*, J. Math. Mech. 17, 59-87 (1967)

148. A.N. Kolmogorov, *Fundamental concepts of probability theory*, Moscow, Nauka, (1974)

149. A. Kossakowski, *On quantum statistical mechanics of non-hamiltonian systems*, Rept. Math. Phys. 3 No. 4, 247-274 (1972)

150. K. Kraus, *General state changes in quantum theory*, Ann. Phys. 64 No. 2, 331-335 (1971)

151. K. Kraus, *States, effects and operations*, Lect. Notes Phys. 190, (1983)

152. B. Kümmerer, *Markov dilations on W^*-algebras*, J. Funct. Anal. 62 No. 2, 139-177 (1985)

153. B. Kümmerer, H. Maassen, *The essentially commutative dilations of dynamical semigroups on M_n*, Comm. Math. Phys. 109, 1-22 (1987)

154. D. S. Lebedev, L. B. Levitin, *The maximal amount of information transmissible by an electromagnetic field*, Information and Control 9, 1-22 (1966)

155. A. Lesniewski, M. B. Ruskai, *Monotone Riemannian metrics and relative entropy on non-commutative probability spaces*, J. Math. Phys. 40, 5702-5724, (1999)

156. E. Lieb, M. B. Ruskai, *Proof of strong subadditivity of quantum mechanical entropy*, J. Math. Phys. 14, 1938-1941 (1973)

157. G. Lindblad, *Entropy, information and quantum measurement*, Comm. Math. Phys. 33, 305-222 (1973)

158. G. Lindblad, *Completely positive maps and entropy inequalities*, Comm. Math. Phys. 40, 147-151 (1975)

159. G. Lindblad, *On the generators of quantum dynamical semigroup*, Comm. Math. Phys. 48, 119-130 (1976)

160. G. Lindblad, *Non-markovian quantum stochastic processes and their entropy*, Comm. Math. Phys. 65, 281-294, (1979)

161. G. Ludwig, *Foundations of quantum mechanics I* , Springer-Verlag, New York-Heidelberg-Berlin, (1983)

162. G.W. Mackey, *Mathematical foundations of quantum mechanics*, W.A. Benjamin Inc., New York, (1963)

163. G.W. Mackey, *Unitary group representations in physics, probability and number theory*, Benjamin Cummings Dubl. Comp., Reading Mass.-London, (1978)

164. J. Manuceau amd A. Verbeure, *Quasifree states of the CCR*, Comm. Math. Phys. **9** No. 4, 293-302 (1968)

165. K. Matsumoto, *A new approach to CR type bound of the pure state model*, LANL e-print quant-ph/9711008 .

166. K. Matsumoto, *Berry's phase in view of quantum estimation theory and its intrinsic relation with the complex structure*, LANL e-print quant-ph/0006076

167. V.P. Maslov, *Operator methods*, Moscow, Nauka, (1973)

168. S. Massar, S. Popescu, *Optimal extraction of information from finite quantum ensembles*, Phys. Rev. Lett. **74**, 1259-1263, (1995)

169. M.B. Menskii, *Path groups, measurements, fields, particles*, Moscow, Nauka, (1983)

170. P.A. Meyer, *Quantum probability for probabilists*, Lect. Notes Math. **1538**, Springer-Verlag, Berlin-Heidelberg-New York, (1993)

171. B. Mielnik, *Global mobility of Schrödinger's particle*, Rep. Math. Phys. **12** No. 3, 331-339 (1977)

172. B. Misra and E.C.G. Sudarshan, *The Zeno's paradox in quantum theory*, J. Math. Phys. **18** No. 4, 756-763 (1977)

173. E. Nelson, *Dynamical theories of Brownian motion*, Princeton, New Jersey, (1967)

174. E. Nelson, *Field theory and the future of stochastic mechanics*, Lect. Notes Phys. **262**, 438-469 (1986)

175. J. von Neumann, *Mathematische Grundlagen der Quantenmechanik*, Springer-Verlag, Berlin, (1932). English translation: *Mathematical foundations of quantum mechanics*, Dover, New York, (1954).

176. M. Nielsen, I. Chuang, *Quantum computation and information*, Cambridge University Press, Cambridge, (2000)

177. M. Ohya, D. Petz, *Quantum entropy and its use*, Lect. Notes Phys. **8**, Springer-Verlag, Berlin (1993)

178. V.I. Oseledets, *Completely positive linear maps, non-Hamiltonian evolution and quantum stochastic processes*, Itogi Nauki Tekhn. Teor. Veroyatn. Mat. Stat. Teor. Kibern VINITI **20** , 52-94 (1983)

179. M. Ozawa, *Optimal measurements for general quantum systems*, Rep. Math. Phys. **18** No. 1, 11-28 (1980)

180. M. Ozawa, *Quantum measuring processes of continuous observables*, J. Math. Phys. **25**, 79-87 (1984)

181. M. Ozawa, *Conditional probability and a posteriori states in quantum mechanics*, Publ. RIMS Kyoto Univ. **21** No. 2, 279-295 (1985)

182. M. Ozawa, *Measuring processes and repeatability hypothesis*, Lect. Notes Math. **1229**, 412-421 (1987)

183. M. Ozawa, *Realization of measurement and the standard quantum limit.* In: *Squeezed and nonclassical light.* Eds. P. Tombesi and E. R. Pike, Plenum Press, New York, pp. 263-286, (1989)

184. K.R. Parthasarathy, *Azema martingales and quantum stochastic calculus.* In: *Proc. R. C. Bose Symp.* Ed. R.R. Bahadur, Wiley Eastern, New Delhi, pp. 551-569, (1990)

185. K.R. Parthasarathy, *An introduction to quantum stochastic calculus*, Birkhäuser Verlag, Basel-Boston-Berlin, (1992)

186. K.R. Parthasarathy and K. Schmidt, *Positive definite kernels, continuous tensor products and central limit theorems of probability theory*, Lect. Notes Math. **272**, (1972)

187. A. Peres, *Quantum theory: concepts and methods*, Kluwer Acad. Publishers, Dordrecht (1993)

188. D. Petz, *Sufficient subalgebras and the relative entropy of states of a von Neumann algebra*, Comm. Math. Phys. **105**, 123-131 (1986)

189. D. Petz, *Monotone metrics on matrix spaces*, Lin. Alg. Appl. **244**, 81-96, (1996)

190. Yu.V. Prokhorov and Yu.A. Rozanov, *Probability theory*, Moscow, Nauka, (1967)

191. *Quantum optics, experimental gravitation and measurement*. Eds. P. Meystre and M.Q. Scully, Plenum Press, New York, (1983)

192. *Quantum communication, computing and measurement* Eds. O. Hirota, A. S. Holevo and C. M. Caves, Plenum Press, New York, (1997)

193. *Quantum probability and applications to the quantum theory of irreversible processes*. Eds. L. Accardi, A. Frigerio and V. Gorini, Lect. Notes Math. **1055**, (1984)

194. *Quantum probability and applications II*. Eds. L. Accardi and W. von Waldenfels, Lect. Notes Math. **1136**, (1985)

195. *Quantum probability and applications III*. Eds. L. Accardi and W. von Waldenfels, Lect. Notes Math. **1303**, (1988)

196. *Quantum probability and applications IV*. Eds. L. Accardi and W. von Waldenfels, Lect. Notes Math. **1396**, (1989)

197. *Quantum probability and applications V* Eds. L. Accardi and W. von Waldenfels, Lect. Notes Math. **1442**, (1990)

198. M. Reed and B. Simon, *Methods of modern mathematical physics II*, Academic Press, New York-San Francisco-London, (1975)

199. F. Riesz and B. Sz-Nagy, *Leçons d'analyse fonctionelle*, 6th edition, (Acad. Kiado, Budapest, 1972)

200. B. L. Rozovskii, *Stochastic evolution systems. Linear theory and applications to non-linear filtering.*, Kluwer, Dordrecht-Boston-London, (1990)

201. T.A. Sarymsakov, *Introduction to quantum probability theory*, FAN, Tashkent, (1985)

202. M. Sasaki, S. M. Barnett, R. Jozsa, M. Osaki, O. Hirota, *Accessible information and optimal strategy for real symmetrical quantum sources*, Phys. Rev. A **59** No. 5 (1999); LANL e-print quant-ph/9812062

203. T.-L. Sauvageot, *Markov quantum semigroups admit covariant Markov C-dilation*, Comm. Math. Phys. **106**, 91-103 (1986)

204. M. Schürmann, *Noncommutative stochastic processes with independent and stationary increments satisfy quantum stochastic differential equations*, Probab. Theory Rel. Fields **84**, 473-490 (1990)

205. H. Scutaru, *Coherent states and induced representations*, Lett. Math. Phys. **2** No. 2, 101-107 (1977)

206. I. Segal, *Mathematical problems of relativistic physics*, Amer. Math. Soc., Providence, Rhode Island, (1963)

207. D. Shale and W. Stinespring, *States of the Clifford algebra*, Ann. of Math. **80** No. 2, 365-381 (1964)

208. A.V. Skorokhod, *Operator stochastic differential equations in stochastic semigroups*, Uspekhi Mat. Nauk **37** No. 6, 157-183 (1982)

209. H. Spohn, *Kinetic equations from Hamiltonian dynamics: Markovian limits*, Rev. Modern Phys. **53** No. 3, 569-615 (1980)

210. M. Srinivas, *Collapse postulate for observables with continuous spectrum*, Comm. Math. Phys. **71**, 131-158 (1980)

211. M. Srinivas and E.B. Davies, *Photon counting probability in quantum optics*, Optica Acta **28** No. 7, 981-996 (1981)

212. A. Steane, *Quantum computing*, Rept. Progress in Physics **61**, 117-173 (1998); LANL e-print quant-ph/9708022

213. R.L. Stratonovich, *The quantum generalization of optimal statistical estimation and hypothesis testing*, Stochastics **1**, 87-126 (1973)

214. R.L. Stratonovich and A.G. Vantsyan, *On asymptotically perfect decoding in pure quantum channels*, Problem. Upr. i Teor. Inform. **7** No. 3, 161-174 (1978)

215. R. F. Streater, *Statistical dynamics: a stochastic approach to nonequilibrium thermodynamics*, Imperial College Press, London, (1995)

216. S.J. Summers and R. Werner, *Bell's inequalities and quantum field theory - II. Bell's inequalities are maximally violated in vacuum*, J. Math. Phys. **28** No. 10, 2448-2456 (1987)

217. M. Takesaki, *Conditional expectations in von Neumann algebra*, J. Funct. Anal. **9**, 306-321 (1972)

218. B.S. Tsirelson, *Quantum analogues of Bell's inequalities - The case of two spatially separated domains*, Zap. Nauchn. Sem. Leningrad Otdel. Math. Inst. Steklov Problems of probabilistic distributions IX **142**, 175-194 (1985)

219. A. Uhlmann, *Density operators as an arena for differential geometry*, Rep. Math. Phys. **33**, 253-263 (1993)

220. H. Umegaki, *Conditional expectations in an operator algebra IV (Entropy and information)*, Kodai Math. Sem. Rep. **14** No. 2, 59-85 (1962)

221. *The uncertainty principle and foundations of quantum mechanics.* Eds. W.C. Price and G.S. Chissick, Wiley, London, (1977)

222. V.S. Varadarajan, *Geometry of quantum theory*, Springer Verlag, New York-Berlin-Heidelberg-Tokyo, (1985)

223. G.F. Vincent-Smith, *Dilation of a dissipative quantum dynamical system to a quantum Markov process*, Proc. London Math. Soc. **59** No. 1, 58-72 (1989)

224. D.F. Walls, G. J. Milburn, *Quantum optics*, Springer-Verlag, Berlin-Heidelberg-New York, (1994)

225. T. Waniewski, *Theorem about completeness of quantum mechanical motion group*, Rept. Math. Phys. **11** No. 3, 331-339 (1977)

226. A. Wehrl, *General properties of entropy*, Rev. Modern Phys. **50**, 221-250 (1978)

227. R. F. Werner, *Quantum harmonic analysis on phase space*, J. Math. Phys. **25**, 1404-1411 (1984)

228. R. F. Werner, *Optimal cloning of pure states*, Phys. Rev. A **A58**, 1827-1832 (1998)

229. H. Weyl, *Gruppentheorie und Quantenmechanik*, (S. Hirzel, Leipzig, 1928). English translation: *The theory of groups and quantum mechanics*, (Dover, New York, 1955)

230. E.P. Wigner, *Symmetries and reflections*, Indiana Univ. Press, Bloomington-London, (1970)

Symbols

Operators, matrices	$A, B, C, \ldots$
Sets of operators	$\mathfrak{A}, \mathfrak{B}, \mathfrak{C}, \ldots$
Real line	$\mathbb{R}$
Complex plane	$\mathbb{C}$
Hilbert spaces	$\mathcal{H}$
Space of self-adjoint operators on $\mathcal{H}$	$\mathfrak{O}(\mathcal{H})$
Algebra of bounded operators on $\mathcal{H}$	$\mathfrak{B}(\mathcal{H})$
Space of trace class operators on $\mathcal{H}$	$\mathfrak{T}(\mathcal{H})$
Space of Hermitian bounded operators on $\mathcal{H}$	$\mathfrak{B}_h(\mathcal{H})$
Space of Hermitian trace class operators on $\mathcal{H}$	$\mathfrak{T}_h(\mathcal{H})$
Logic of projection operators on $\mathcal{H}$	$\mathfrak{E}(\mathcal{H})$
Set of density operators on $\mathcal{H}$	$\mathfrak{S}(\mathcal{H})$
Sets of generalized observables	$\mathfrak{M}$
Generators and operations	$\mathfrak{L}, \mathfrak{M}, \ldots$
σ-algebra of Borel subsets of a set $\mathcal{X}$	$\mathfrak{B}(\mathcal{X})$
Space of square-integrable functions on $\mathcal{X}$	$L^2(\mathcal{X})$
Space of continuous functions on $\mathcal{X}$	$C(\mathcal{X})$
Real part	$\mathfrak{R}$
Imaginary part	$\mathfrak{I}$
Mean value of an observable X in the state S	$\mathbf{E}_S(X)$
Variance of an observable X in the state S	$\mathbf{D}_S(X)$

Index

n-particle subspace, 119
*-homomorphism, 72
*-weak convergence, 14

accretive operator, 86
adapted process, 122
adjoint operator, 14
annihilation operator, 120
approximately localizable system, 70

backward Markov master equation, 86
Bayes decision rule, 46, 54
Bayes risk, 46
Bell-Clauser-Horn-Shimony (BCHS)
   inequality, 35
Bures metric, 59

C^*-algebra, 71
canonical commutation relations, 24
canonical observables, 26
capacity, 51
center, 4
characteristic function, 28, 111
cocycle, 85, 126
coherent states, 28
commutant, 76
commutator, 18
compatible observables, 4, 18
complete positivity, 11
completely positive map, 71, 98
conditionally completely positive map,
   84
conservative stochastic differential
   equation, 144
conservativity condition, 86
contextual hidden variables, 33
convolution semigroup of instruments,
   114
correlation kernels, 94
correlations, 20
counting process, 117
covariance, 19

covariant dynamical semigroup, 88
covariant observable, 61
creation operator, 120

decision rule, 45, 54
density matrix, 1
density operator, 15
detailed balance condition, 94
deterministic observable, 44
deterministic transition probabilities,
   43
dilation, 11
Dirac notation, 13
dissipative stochastic equation, 143
duality map, 131
dynamical map, 10, 74
dynamical semigroup, 11

Einstein locality, 34
entangled vector, 33
entropy, 15
estimate, 55
exponential vector, 121
extreme point, 15

Fock space, 29
form-generator, 86
forward Markov master equation, 86
functionally subordinate, 4

gauge transformations, 66
Gaussian state, 28
Gelfand-Naimark-Segal (GNS)
   construction, 72
generalized coherent states, 64
generalized observable, 41
generalized statistical model of
   quantum mechanics, 6, 41
generator, 116
Gleason's theorem, 16
ground state, 27

H-theorem, 74
Hamiltonian, 24
Heisenberg picture, 22
Heisenberg uncertainty relation, 27
Hermitian operator, 14
hidden variables, 5
hyperdissipativity, 88

imprimitivity system, 64
infinitely divisible instrument, 111
information openness, 7
instrument, 9, 97
instrumental process with independent
    increments (i-process), 115
irreversibility, 10
isometric operator, 14

Jordan (symmetrized) product, 20

Kadison-Schwarz inequality, 71
Kolmogorov model, 4
Kolmogorov-Daniel construction, 11

localizability, 8
localizable system, 70
locally unbiased decision rule, 57

Markov process, 94
martingale, 122
mean value, 17
minimal uncertainty states, 27
minimax decision rule, 54
mixture, 3
momentum, 24

nondemolition measurement, 108
nonseparability, 5
normal map, 73
normal operator, 71
normal state, 18

observable, 1, 41
open system, 10
operation, 73
orthogonal resolution of the identity, 16
overcomplete family, 40

partial trace, 31
particle number, 120
Pauli matrices, 20
position observable, 25
positive map, 71
positive operator, 14
posterior states, 98

principle of complementarity, 5
probability distribution of an observ-
    able, 3
process, 122
projection, 14
projection postulate, 99
pure state, 17

quantum communication channel, 7
quantum dynamical semigroup, 81
quantum entanglement, 8
quantum events, 2
quantum logic, 2
quantum stochastic calculus, 11
quantum stochastic process, 93
quantum-representable matrices, 36
quasi-free states, 28
quasicharacteristic function, 114
qubit, 20

real observable, 17
relative entropy, 74
repeatability hypothesis, 103
repeatable instrument, 103
reproducibility, 3
resolution of the identity, 6, 39
right, left logarithmic derivatives, 58

Schrödinger picture, 22
Schrödinger representation, 25
selfadjoint operator, 16
separability, 34
separated statistical model, 4
Shannon information, 46
simplex, 1
spectral measure, 16
spectral theorem, 16
squeezed states, 28
stability of frequencies, 3
standard measurable space, 39
standard statistical model, 3
standard statistical model of quantum
    mechanics, 17
state, 1, 17, 18
stationary Markov process, 94
statistical ensemble, 3
statistical model, 4
Stone-von Neumann uniqueness
    theorem, 26
strong convergence, 14
symmetric Fock space, 119
symmetrized logarithmic derivatives, 57
symmetry, 22
symplectic, 26

tensor product, 29
time-ordered exponential, 128
trace, 14
trace class operator, 14
trace norm, 14

unbiased decision rule, 56
uncertainty relation, 19
unitary operator, 14

vacuum vector, 121
variance, 19
velocity observable, 25
von Neumann algebra, 18
von Neumann projection postulate, 9

Wald model, 6
weak convergence, 14

Printing: Weihert-Druck GmbH, Darmstadt
Binding: Buchbinderei Schäffer, Grünstadt

Lecture Notes in Physics

For information about Vols. 1–531
please contact your bookseller or Springer-Verlag

Vol. 532: S. C. Müller, J. Parisi, W. Zimmermann (Eds.), Transport and Structure. Their Competitive Roles in Biophysics and Chemistry. XII, 400 pages. 1999.

Vol. 533: K. Hutter, Y. Wang, H. Beer (Eds.), Advances in Cold-Region Thermal Engineering and Sciences. Proceedings, 1999. XIV, 608 pages. 1999.

Vol. 534: F. Moreno, F. González (Eds.), Light Scattering from Microstructures. Proceedings, 1998. XII, 300 pages. 2000

Vol. 535: H. Dreyssé (Ed.), Electronic Structure and Physical Properties of Solids: The Uses of the LMTO Method. Proceedings, 1998. XIV, 458 pages. 2000.

Vol. 536: T. Passot, P.-L. Sulem (Eds.), Nonlinear MHD Waves and Turbulence. Proceedings, 1998. X, 385 pages. 1999.

Vol. 537: S. Cotsakis, G. W. Gibbons (Eds.), Mathematical and Quantum Aspects of Relativity and Cosmology. Proceedings, 1998. XII, 251 pages. 1999.

Vol. 538: Ph. Blanchard, D. Giulini, E. Joos, C. Kiefer, I.-O. Stamatescu (Eds.), Decoherence: Theoretical, Experimental, and Conceptual Problems. Proceedings, 1998. XII, 345 pages. 2000.

Vol. 539: A. Borowiec, W. Cegła, B. Jancewicz, W. Karwowski (Eds.), Theoretical Physics. Fin de Siècle. Proceedings, 1998. XX, 319 pages. 2000.

Vol. 540: B. G. Schmidt (Ed.), Einstein's Field Equations and Their Physical Implications. Selected Essays. 1999. XIII, 429 pages. 2000.

Vol. 541: J. Kowalski-Glikman (Ed.), Towards Quantum Gravity. Proceedings, 1999. XII, 376 pages. 2000.

Vol. 542: P. L. Christiansen, M. P. Sørensen, A. C. Scott (Eds.), Nonlinear Science at the Dawn of the 21st Century. Proceedings, 1998. XXVI, 458 pages. 2000.

Vol. 543: H. Gausterer, H. Grosse, L. Pittner (Eds.), Geometry and Quantum Physics. Proceedings, 1999. VIII, 408 pages. 2000.

Vol. 544: T. Brandes (Ed.), Low-Dimensional Systems. Interactions and Transport Properties. Proceedings, 1999. VIII, 219 pages. 2000

Vol. 545: J. Klamut, B. W. Veal, B. M. Dabrowski, P. W. Klamut, M. Kazimierski (Eds.), New Developments in High-Temperature Superconductivity. Proceedings, 1998. VIII, 275 pages. 2000.

Vol. 546: G. Grindhammer, B. A. Kniehl, G. Kramer (Eds.), New Trends in HERA Physics 1999. Proceedings, 1999. XIV, 460 pages. 2000.

Vol. 547: D. Reguera, G. Platero, L. L. Bonilla, J. M. Rubí(Eds.), Statistical and Dynamical Aspects of Mesoscopic Systems. Proceedings, 1999. XII, 357 pages. 2000.

Vol. 548: D. Lemke, M. Stickel, K. Wilke (Eds.), ISO Surveys of a Dusty Universe. Proceedings, 1999. XIV, 432 pages. 2000.

Vol. 549: C. Egbers, G. Pfister (Eds.), Physics of Rotating Fluids. Selected Topics, 1999. XVIII, 437 pages. 2000.

Vol. 550: M. Planat (Ed.), Noise, Oscillators and Algebraic Randomness. Proceedings, 1999. VIII, 417 pages. 2000.

Vol. 551: B. Brogliato (Ed.), Impacts in Mechanical Systems. Analysis and Modelling. Lectures, 1999. IX, 273 pages. 2000.

Vol. 552: Z. Chen, R. E. Ewing, Z.-C. Shi (Eds.), Numerical Treatment of Multiphase Flows in Porous Media. Proceedings, 1999. XXI, 445 pages. 2000.

Vol. 553: J.-P. Rozelot, L. Klein, J.-C. Vial Eds.), Transport of Energy Conversion in the Heliosphere. Proceedings, 1998. IX, 214 pages. 2000.

Vol. 554: K. R. Mecke, D. Stoyan (Eds.), Statistical Physics and Spatial Statistics. The Art of Analyzing and Modeling Spatial Structures and Pattern Formation. Proceedings, 1999. XII, 415 pages. 2000.

Vol. 555: A. Maurel, P. Petitjeans (Eds.), Vortex Structure and Dynamics. Proceedings, 1999. XII, 319 pages. 2000.

Vol. 556: D. Page, J. G. Hirsch (Eds.), From the Sun to the Great Attractor. X, 330 pages. 2000.

Vol. 557: J. A. Freund, T. Pöschel (Eds.), Stochastic Processes in Physics, Chemistry, and Biology. X, 330 pages. 2000.

Vol. 558: P. Breitenlohner, D. Maison (Eds.), Quantum Field Theory. Proceedings, 1998. VIII, 323 pages. 2000

Vol. 559: H.-P. Breuer, F. Petruccione (Eds.), Relativistic Quantum Measurement and Decoherence. Proceedings, 1999. X, 140 pages. 2000.

Vol. 560: S. Abe, Y. Okamoto (Eds.), Nonextensive Statistical Mechanics and Its Applications. IX, 272 pages. 2001.

Vol. 561: H. J. Carmichael, R. J. Glauber, M. O. Scully (Eds.), Directions in Quantum Optics. XVII, 369 pages. 2001.

Vol. 562: C. Lämmerzahl, C. W. F. Everitt, F. W. Hehl (Eds.), Gyros, Clocks, Interferometers...: Testing Relativistic Gravity in Space. XVII,507 pages. 2001.

Vol. 563: F. C. Lázaro, M. J. Arévalo (Eds.), Binary Stars. Selected Topics on Observations and Physical Processes. 1999. X, 332 pages. 2001.

Vol. 564: T. Pöschel, S. Luding (Eds.), Granular Gases. VIII, 457 pages. 2001.

Vol. 565: E. Beaurepaire, F. Scheurer, G. Krill, J.-P. Kappler (Eds.), Magnetism and Synchrotron Radiation. XIV, 396 pages. 2001.

Vol. 566: J. L. Lumley (Ed.), Fluid Mechanics and the Environment: Dynamical Approaches. VIII, 412 pages. 2001.

Vol. 567: D. Reguera, L. L. Bonilla, J. M. Rubí (Eds.), Coherent Structures in Complex Systems. IX, 465 pages. 2001.

Vol. 568: P. A. Vermeer, S. Diebels, W. Ehlers, H. J. Herrmann, S. Luding, E. Ramm (Eds.), Continuous and Discontinuous Modelling of Cohesive-Frictional Materials. XIV, 307 pages. 2001.

Monographs
For information about Vols. 1–24
please contact your bookseller or Springer-Verlag

Vol. m 25: A. V. Bogdanov, G. V. Dubrovskiy, M. P. Krutikov, D. V. Kulginov, V. M. Strelchenya, Interaction of Gases with Surfaces. XIV, 132 pages. 1995.

Vol. m 26: M. Dineykhan, G. V. Efimov, G. Ganbold, S. N. Nedelko, Oscillator Representation in Quantum Physics. IX, 279 pages. 1995.

Vol. m 27: J. T. Ottesen, Infinite Dimensional Groups and Algebras in Quantum Physics. IX, 218 pages. 1995.

Vol. m 28: O. Piguet, S. P. Sorella, Algebraic Renormalization. IX, 134 pages. 1995.

Vol. m 29: C. Bendjaballah, Introduction to Photon Communication. VII, 193 pages. 1995.

Vol. m 30: A. J. Greer, W. J. Kossler, Low Magnetic Fields in Anisotropic Superconductors. VII, 161 pages. 1995.

Vol. m 31 (Corr. Second Printing): P. Busch, M. Grabowski, P.J. Lahti, Operational Quantum Physics. XII, 230 pages. 1997.

Vol. m 32: L. de Broglie, Diverses questions de mécanique et de thermodynamique classiques et relativistes. XII, 198 pages. 1995.

Vol. m 33: R. Alkofer, H. Reinhardt, Chiral Quark Dynamics. VIII, 115 pages. 1995.

Vol. m 34: R. Jost, Das Märchen vom Elfenbeinernen Turm. VIII, 286 pages. 1995.

Vol. m 35: E. Elizalde, Ten Physical Applications of Spectral Zeta Functions. XIV, 224 pages. 1995.

Vol. m 36: G. Dunne, Self-Dual Chern-Simons Theories. X, 217 pages. 1995.

Vol. m 37: S. Childress, A.D. Gilbert, Stretch, Twist, Fold: The Fast Dynamo. XI, 406 pages. 1995.

Vol. m 38: J. González, M. A. Martín-Delgado, G. Sierra, A. H. Vozmediano, Quantum Electron Liquids and High-Tc Superconductivity. X, 299 pages. 1995.

Vol. m 39: L. Pittner, Algebraic Foundations of Non-Com-mutative Differential Geometry and Quantum Groups. XII, 469 pages. 1996.

Vol. m 40: H.-J. Borchers, Translation Group and Particle Representations in Quantum Field Theory. VII, 131 pages. 1996.

Vol. m 41: B. K. Chakrabarti, A. Dutta, P. Sen, Quantum Ising Phases and Transitions in Transverse Ising Models. X, 204 pages. 1996.

Vol. m 42: P. Bouwknegt, J. McCarthy, K. Pilch, The W₃ Algebra. Modules, Semi-infinite Cohomology and BV Algebras. XI, 204 pages. 1996.

Vol. m 43: M. Schottenloher, A Mathematical Introduction to Conformal Field Theory. VIII, 142 pages. 1997.

Vol. m 44: A. Bach, Indistinguishable Classical Particles. VIII, 157 pages. 1997.

Vol. m 45: M. Ferrari, V. T. Granik, A. Imam, J. C. Nadeau (Eds.), Advances in Doublet Mechanics. XVI, 214 pages. 1997.

Vol. m 46: M. Camenzind, Les noyaux actifs de galaxies. XVIII, 218 pages. 1997.

Vol. m 47: L. M. Zubov, Nonlinear Theory of Dislocations and Disclinations in Elastic Body. VI, 205 pages. 1997.

Vol. m 48: P. Kopietz, Bosonization of Interacting Fermions in Arbitrary Dimensions. XII, 259 pages. 1997.

Vol. m 49: M. Zak, J. B. Zbilut, R. E. Meyers, From Instability to Intelligence. Complexity and Predictability in Nonlinear Dynamics. XIV, 552 pages. 1997.

Vol. m 50: J. Ambjørn, M. Carfora, A. Marzuoli, The Geometry of Dynamical Triangulations. VI, 197 pages. 1997.

Vol. m 51: G. Landi, An Introduction to Noncommutative Spaces and Their Geometries. XI, 200 pages. 1997.

Vol. m 52: M. Hénon, Generating Families in the Restricted Three-Body Problem. XI, 278 pages. 1997.

Vol. m 53: M. Gad-el-Hak, A. Pollard, J.-P. Bonnet (Eds.), Flow Control. Fundamentals and Practices. XII, 527 pages. 1998.

Vol. m 54: Y. Suzuki, K. Varga, Stochastic Variational Approach to Quantum-Mechanical Few-Body Problems. XIV, 324 pages. 1998.

Vol. m 55: F. Busse, S. C. Müller, Evolution of Spontaneous Structures in Dissipative Continuous Systems. X, 559 pages. 1998.

Vol. m 56: R. Haussmann, Self-consistent Quantum Field Theory and Bosonization for Strongly Correlated Electron Systems. VIII, 173 pages. 1999.

Vol. m 57: G. Cicogna, G. Gaeta, Symmetry and Perturbation Theory in Nonlinear Dynamics. XI, 208 pages. 1999.

Vol. m 58: J. Daillant, A. Gibaud (Eds.), X-Ray and Neutron Reflectivity: Principles and Applications. XVIII, 331 pages. 1999.

Vol. m 59: M. Kriele, Spacetime. Foundations of General Relativity and Differential Geometry. XV, 432 pages. 1999.

Vol. m 60: J. T. Londergan, J. P. Carini, D. P. Murdock, Binding and Scattering in Two-Dimensional Systems. Applications to Quantum Wires, Waveguides and Photonic Crystals. X, 222 pages. 1999.

Vol. m 61: V. Perlick, Ray Optics, Fermat's Principle, and Applications to General Relativity. X, 220 pages. 2000.

Vol. m 62: J. Berger, J. Rubinstein, Connectivity and Superconductivity. XI, 246 pages. 2000.

Vol. m 63: R. J. Szabo, Ray Optics, Equivariant Cohomology and Localization of Path Integrals. XII, 315 pages. 2000.

Vol. m 64: I. G. Avramidi, Heat Kernel and Quantum Gravity. X, 143 pages. 2000.

Vol. m 65: M. Hénon, Generating Families in the Restricted Three-Body Problem. Quantitative Study of Bifurcations. XII, 301 pages. 2001.

Vol. m 66: F. Calogero, Classical Many-Body Problems Amenable to Exact Treatments. XIX, 749 pages. 2001.

Vol. m 67: A. S. Holevo, Statistical Structure of Quantum Theory. IX, 159 pages. 2001.

Printed in the United States
87001LV00003B/307-312/A